Synthesis Lectures on Engineering, Science, and Technology

The focus of this series is general topics, and applications about, and for, engineers and scientists on a wide array of applications, methods and advances. Most titles cover subjects such as professional development, education, and study skills, as well as basic introductory undergraduate material and other topics appropriate for a broader and less technical audience.

Vladimir Afonso · Murilo Perleberg ·
Bruno Zatt · Marcelo Porto · Luciano Agostini ·
Altamiro Susin

Hardware Design for 3D Video Coding

Energy-efficient and High-throughput
Architectures for Multiview Plus Depth
Video Content

 Springer

Vladimir Afonso
Video Technology Research Group
Instituto Federal Sul-rio-grandense
Pelotas, Rio Grande do Sul, Brazil

Bruno Zatt
Video Technology Research Group
Universidade Federal de Pelotas
Pelotas, Rio Grande do Sul, Brazil

Luciano Agostini
Video Technology Research Group
Universidade Federal de Pelotas
Pelotas, Rio Grande do Sul, Brazil

Murilo Perleberg
Video Technology Research Group
Universidade Federal de Pelotas
Pelotas, Rio Grande do Sul, Brazil

Marcelo Porto
Video Technology Research Group
Universidade Federal de Pelotas
Pelotas, Rio Grande do Sul, Brazil

Altamiro Susin
Microelectronics Research Group
Federal University of Rio Grande do Sul
Porto Alegre, Rio Grande do Sul, Brazil

ISSN 2690-0300   ISSN 2690-0327 (electronic)
Synthesis Lectures on Engineering, Science, and Technology
ISBN 978-3-031-80231-7   ISBN 978-3-031-80232-4 (eBook)
https://doi.org/10.1007/978-3-031-80232-4

This Springer imprint is published by the registered company Springer Nature Switzerland AG
The registered company address is: Gewerbestrasse 11, 6330 Cham, Switzerland

If disposing of this product, please recycle the paper.

Preface

The intensive use of multimedia content in our daily tasks, including work, study, and entertainment, has driven the development of increasingly immersive visual information technologies. Among these are omnidirectional videos, three-dimensional (3D) videos, virtual reality, augmented reality, mixed reality, light fields, point clouds, and holograms. Each of these technologies is at a different stage of development, but it is expected that shortly, highly immersive multimedia content will become widely available, alongside devices capable of supporting such content.

However, the widespread adoption of these technologies still faces significant challenges. In part, due to the excessive computational costs required to process such content. Additionally, there is the difficulty of creating devices that can support and display this content comfortably for users. Nevertheless, the industry and academia are making considerable efforts to overcome these obstacles. This book is a contribution in this direction, with a focus on 3D videos using the multiview plus depth (MVD) format.

Three-dimensional videos are captured by a set of cameras, all recording the same scene but from slightly different positions. By combining these different video streams, it becomes possible to convey a sense of depth to the viewer, enhancing their immersive experience with the presented content. The MVD format emerged as an alternative to represent 3D videos more efficiently, offering greater flexibility in the immersive experience. In this case, in addition to the texture information captured by conventional cameras, a depth map of the scene is also captured. This depth map is a grayscale image that indicates the distance of objects in the scene relative to the camera. With the texture and depth map information, it is possible, on the decoder side, to generate synthetic texture views (which were not captured) between the original captured views, thereby improving the quality of the immersive experience.

Naturally, the use of MVD technology presents its own set of challenges that need to be overcome. First, multiple texture and depth channels must be captured, encoded, transmitted, and/or stored. In other words, the amount of data required to represent an MVD video is significantly greater than that needed to represent a two-dimensional video.

With this much larger data volume to process, the computational effort required to encode these videos is substantially increased. Moreover, to enhance the coding efficiency of such content, new coding tools have been proposed, further increasing the computational cost of both the encoding and decoding processes.

To address these challenges, the use of dedicated hardware is essential. Additionally, it is crucial to explore new algorithms, or simplified versions of existing ones, to reduce the computational cost and memory bandwidth required by the coding tools.

This book contributes to the current efforts in this area, presenting dedicated algorithms and architectures for 3D video coding, specifically considering MVD content, with a focus on 3D High-Efficiency Video Coding (3D-HEVC) as a case study. 3D-HEVC is an extension of the HEVC standard and it was developed by the Joint Collaborative Team on 3D Video Coding Extension Development (JCT-3V), a group of ITU-T and ISO experts.

Initially, basic concepts regarding MVD video coding are introduced. Subsequently, the 3D extension of the High-Efficiency Video Coding (HEVC) standard, referred to as 3D-HEVC, is presented, along with a discussion of its principal tools.

A comprehensive experimental evaluation of 3D-HEVC is then provided, utilizing the 3D-HTM reference software. These evaluations are crucial for a deeper understanding of the behavior of 3D-HEVC and its key tools, and they have been instrumental in the design of the hardware to be presented in this book. Assessments of time profiling and memory accesses for intra-frame, inter-frame, and inter-view predictions are initially discussed. Following this, detailed evaluations of the 3D-HEVC tools are presented, encompassing their behavior in processing both texture information and depth map data. These assessments consider both the Conventional Coding Order (CCO) configuration and the Flexible Coding Order (FCO) configuration, which has been explored in the hardware developed in this book.

Next, dedicated hardware architectures for various 3D-HEVC tools are presented, along with the heuristics developed to enhance hardware design efficiency in terms of processing rate, area, and power dissipation. Various architectures for intra-frame, inter-frame, and inter-view predictions are introduced, along with their respective heuristics.

For intra-frame prediction, a low-power architecture for the Depth Intra Skip (DIS) tool is presented, accompanied by its corresponding heuristic. Additionally, a low-power architecture with reduced memory bandwidth for the entire intra-frame prediction system is introduced. This solution is also based on heuristics that are presented and evaluated. Furthermore, a low-power architecture for the Depth Modeling Mode 1 (DMM-1) tool is discussed, including its associated heuristic. Finally, a comprehensive architecture for intra-frame prediction of 3D-HEVC, exploring the FCO configuration of 3D-HEVC, is presented. All previously mentioned architectures were designed for the CCO configuration of 3D-HEVC. The latter architecture is configurable, featuring various operational points relating to quality and power consumption. The architecture utilizing the FCO employs multiple heuristics, which are thoroughly explained and evaluated.

Subsequently, the architectures developed for inter-frame and inter-view predictions are presented. Initially, an architecture for the entire inter-frame and inter-view prediction system is introduced, emphasizing low power consumption and a runtime adaptive memory hierarchy. Several sets of constraints have been evaluated to define a heuristic that enables the design of low-power hardware. These constraints are presented and evaluated, with the most promising combinations selected for the overall hardware system design. Given that inter-frame and inter-view predictions encompass numerous coding tools with significant computational costs and high memory bandwidth requirements, heuristic algorithms have been defined for each of the main tools involved in these predictions. The next architecture presented is a low-power architecture specifically designed for disparity estimation in inter-view prediction. Two novel disparity search heuristics are introduced, aimed at exploiting the displacement between cameras to reduce computational costs. The heuristic yielding the best results has been implemented in hardware, and its architecture is discussed in detail.

From the foregoing, it is possible to conclude that this book presents, in a single volume:

(a) Detailed information on 3D video coding, considering MVD content, with 3D-HEVC serving as a case study.
(b) A comprehensive set of experimental results evaluating the coding of 3D content using MVD, with a focus on the 3D-HEVC reference software and its primary coding tools.
(c) Various hardware-friendly heuristics for the 3D-HEVC, including the reduction of computational costs and memory bandwidth while minimizing impacts on coding efficiency.
(d) Several innovative hardware architectures for the most computationally intensive stages of 3D-HEVC: intra-frame, inter-frame, and inter-view predictions, with an emphasis on low power dissipation and low energy consumption.

Thus, this book offers a complete and unprecedented set of contributions to the field of 3D video coding based on MVD content. Although the developed heuristics and dedicated hardware were focused on the 3D-HEVC, the proposed solutions address the primary challenges associated with coding this type of visual content, then these solutions can be adapted to other 3D video encoders.

We hope that readers find the content of this book relevant to their work in the field and can leverage the ideas presented for future projects.

Pelotas, Brazil   Vladimir Afonso
Pelotas, Brazil   Murilo Perleberg
Pelotas, Brazil   Bruno Zatt
Pelotas, Brazil   Marcelo Porto
Pelotas, Brazil   Luciano Agostini
Porto Alegre, Brazil   Altamiro Susin
October 2024

Acknowledgments

This work was partly funded by the Coordenação de Aperfeiçoamento de Pessoal de Nível Superior—Brasil (CAPES). It was also financed in part by the Fundação de Amparo à pesquisa do Estado do Rio Grande do Sul—Brasil (FAPERGS), and by the Conselho Nacional de Desenvolvimento Científico e Tecnológico—Brasil (CNPq).

This work was also partly supported by the Graduate Program on Microelectronics (PGMicro), Institute of Informatics, Federal University of Rio Grande do Sul (UFRGS), and by the Graduate Program in Computing (PPGC), Center for Technological Development (CDTec), Federal University of Pelotas (UFPel).

Contents

Abbreviations and Acronyms

2D	Two-Dimensional
3D	Three-Dimensional
3D-HEVC	3D-High-Efficiency Video Coding
3D-HTM	3D-HEVC Test Model
3D-TV	Three-Dimensional Television
6DoF	Six Degrees of Freedom
6WR	Six Wedgelets and Six Refinements
AI	All Intra
AR	Augmented Reality
ASIC	Application-Specific Integrated Circuit
ASW	Adaptive Search Window
AU	Access Unit
AVC	Advanced Video Coding
BD	Bjontegaard Difference
BD-BR	Bjontegaard Difference Bit Rate
BD-PSNR	Bjontegaard Difference Peak Signal-to-Noise Ratio
BD-Rate	Bjontegaard Difference Bit Rate
B-Encoder	Baseline Encoder
B-Frames	Bi-predicted Frames
BMA	Block Matching Algorithm
BMH	Base Memory Hierarchy
BSM	Binary Segmentation Mask
BV	Base View
CABAC	Context-Based Adaptive Binary Arithmetic Coding
CB	Coding Block
CCO	Conventional Coding Order
CPU	Central Processing Unit
CPV	Constant Partition Value

CTC	Common Test Conditions of 3DV Core Experiments
CTU	Coding Tree Unit
CU	Coding Unit
dB	Decibel
DBBP	Depth-Based Block Partitioning
DCP	Disparity-Compensated Prediction
DE	Disparity Estimation
DIBR	Depth-Image-Based Rendering
DIS	Depth Intra Skip
DMM	Depth Modeling Modes
DMMFast	Depth Modeling Modes Fast Prediction
DPB	Decoded Picture Buffer
DSWR	Depth-Based Dynamic Search Window Resizing
DV	Dependent View
DVS	Dynamic Voltage Scaling
EWS	Explicit Wedgelet Signaling
FCO	Flexible Coding Order
FIR	Finite Impulse Response
FME	Fractional Motion Estimation
FPGA	Field Programmable Gate Array
fps	Frames per second
FS	Full Search
FTV	Free Viewpoint Television
GoP	Group of Pictures
GPB	Generalized P and B Picture
GPU	Graphics Processing Unit
HC	Hardware-oriented Constraint
HD	High Definition
HDS	Horizontal Disparity Search
HEVC SC	HEVC Simulcast
HEVC	High-Efficiency Video Coding
HM	HEVC Test Model
HOTZS	Hardware-Oriented Test Zone Search
ICDSD	Inter-Channel Directional Structure Detector
ICPCP	Inter-Component-Predicted Contour Partitioning
IDR	Instantaneous Decoding Refresh
IEC	International Electrotechnical Commission
IME	Integer Motion Estimation
IO	Intra Only
IPH	Horizontal Intra Prediction
IPHOC	Intra-Prediction Hardware-Oriented Constraints

IPV	Vertical Intra Prediction
ISO	International Organization for Standardization
ITU-T	International Telecommunication Union—Telecommunication Standardization Sector
iUDS	Improved Unidirectional Disparity Search
JCT-3V	Joint Collaborative Team on 3D Video Coding Extension Development
JCT-VC	Joint Collaborative Team on Video Coding
LC-ICDSD	Low Complexity Inter-Channel Directional Structure Detector
LD	Low Delay
MC	Motion Compensation
MCL	Merge Candidate List
MCP	Motion-Compensated Prediction
ME	Motion Estimation
MIV	MPEG Immersive Video
MPEG	Moving Picture Experts Group
MPM	Most Probable Mode
MR	Mixed Reality
MSB	Most Significant Bit
MSE	Mean Squared Error
MVC	Multi-view Video Coding
MVD	Multi-view Video plus Depth
MV-HEVC	Multi-view HEVC
PB	Prediction Block
PC	Processing Core
PS	Power State
PSNR	Peak Signal-to-Noise Ratio
PU	Prediction Unit
QP	Quantization Parameter
quadtree	Quaternary tree
RA	Random Access
RAH	Run-Time Adaptive Hierarchy
RD	Rate-Distortion
RDO	Rate-Distortion Optimization
ROM	Read Only Memory
RSH	Reduced-Size Hierarchy
SA	Search Area
SAD	Sum of Absolute Differences
SAO	Sample Adaptive Offset
SATD	Sum of Absolute Transformed Differences
SC	Sleep Circuitry
SCU	Search and Comparison Unit

SD	Single Depth
SDC	Segment-wise DC Coding
SDH	Horizontal Single Depth
SDV	Vertical Single Depth
SED	Simplified Edge Detection
SR	Search Range
SRAM	Static Random Access Memory
SSD	Sum of Squared Differences
SVDC	Synthesized View Distortion Change
SW	Search Window
TC	Test Case
TMVI	Texture-Based Motion Vector Inheritance
TQ	Transform/Quantization
TU	Transform Unit
TV	Television
TZS	Test Zone Search
UDS	Unidirectional Disparity Search
UHD	Ultra High Definition
VCEG	Video Coding Experts Group
VHDL	VHSIC Hardware Description Language
VHSIC	Very High-Speed Integrated Circuit
VLSI	Very Large Scale Integration
VR	Virtual Reality
VSO	View Synthesis Optimization
YCbCr	Luminance, Chrominance blue, Chrominance red

List of Figures

List of Tables

Introduction

Immersive video technologies, such as 3D-movies, living games, virtual reality, augmented reality, mixed reality (among other technologies) are advancing along with the development of handheld devices. Concurrently, there is a massive popularization of multimedia services and video-capable mobile devices, such as smartphones and tablets. This popularization is pushed by the concern of the users in high-quality and immersive video-related experiences, and by the platforms with support for video sharing, such as social networks (Facebook, Instagram, TikTok, etc.), messaging platforms (WhatsApp, Telegram, Messenger, Viber, etc.), and video-streaming services (Netflix, YouTube, Amazon Prime Video, etc.).

Three-dimensional (3D) videos allow the spectators to perceive the distance (or depth) of the objects in a scene, with a more realistic representation than using the conventional 2D (two-dimensional) videos. In the last decades, the industry has invested in the design and manufacturing of electronic devices with support for recording and reproducing 3D videos. In 2009 the success of the 3D movie Avatar in the cinema (Sony 2024) was a strong booster for this investment. At that time, 3D TVs, 3D Blu-Ray players, and portable devices, such as smartphones, gaming consoles, and camcorders (Toshiba 2024; Lg 2024; Nintendo 2024; Panasonic 2024)—usually based on a stereoscopic approach—were heavily commercialized. However, over the years, the shortage of high-quality 3D-video content, and some limitations presented by the technology, such as the huge amount of data to be transmitted, limited view synthesis performance and the need for a pair of glasses (major part of technologies), led the sales to a slowing down due to the visual discomfort and poor user experience (Pereira et al. 2016). Despite the effort to provide an immersive experience, the first generation of devices capable of dealing with 3D videos

V. Afonso et al., *Hardware Design for 3D Video Coding*, Synthesis Lectures on Engineering, Science, and Technology, https://doi.org/10.1007/978-3-031-80232-4_1

has failed, and the interruption of 3D-TV production by the major manufacturers occurred in 2016 (Lifewire 2024).

As an alternative to some of the problems presented by the previous 3D-video technology, the MVD (Multi-View plus Depth) concept emerged along with 3D-HEVC (3D High Efficiency Video Coding) (Tech et al. 2016)—the 3D extension of the HEVC (High Efficiency Video Coding) standard (Sullivan et al. 2012). Due to the huge amount of data to be processed, stored, and transmitted, the processing of digital videos with increased resolutions, considering 3D videos, is unfeasible without the use of novel and efficient compression techniques in addition to the ones normally applied in the HEVC and in other previous standards. The HEVC was published in April 2013 (ITU-T 2015), initially developed to encode 2D videos. The 3D-HEVC development was finished in February 2015 (ITU-T 2015), with the main purpose of making feasible efficient 3D-video storage and transmission for multi-view systems, such as Free-viewpoint Television (FTV) and Three-dimensional Television (3D-TV), besides supporting emerging technologies such as multi-view auto-stereoscopic displays that do not require the usage of glasses (Dimenco 2024).

The MVD approach is the technology applied in the current and future generations of devices capable of dealing with immersive videos. One important example is the novel MPEG Immersive Video (MIV), the current state-of-the-art standard for immersive technology compression, which was defined using the MVD approach (Boyce et al. 2021). The 3D content is generated using multiple cameras capturing the same scene, but at a little different point of views. These cameras generate the multi-view content allowing, after a necessary processing, the immersive experience to the spectators. The MVD format introduces the concept of depth maps to increase the flexibility of the immersive experience. Then, the multi-view cameras must be able to generate the textures (2D pictures conventionally shown to the spectators that contain the brightness and color information) and, also the depth maps. The depth maps are represented as grayscale pictures that depict the distance between each point in the scene and the camera (Kauff et al. 2007). The objects farthest from the camera are represented by darker shades of gray, while objects (or part of objects) nearest to the camera are represented by light shades of gray. For each video view in MVD format, there is a depth map in addition to the texture picture. The depth maps are not directly visualized by the spectators, but they are used together with the captured texture views at the decoder side to generate additional synthesized texture views through DIBR (Depth Image-Based Rendering) techniques (Kauff et al. 2007), improving the immersive experience with a limited number of captured views, which is important to reduce the amount of data to be captured, processed, and transmitted.

Thus, considering an MVD video system, the transmitter encodes and sends a reduced set of views in comparison to previous texture-only approaches, such as the MVC (Multiview Video Coding) (ITU-T 2008), for obtaining the same number of views after the decoding process, since the receiver decodes the transmitted views and synthesizes virtual views when necessary. Figure 1.1 shows an example of two texture pictures and their

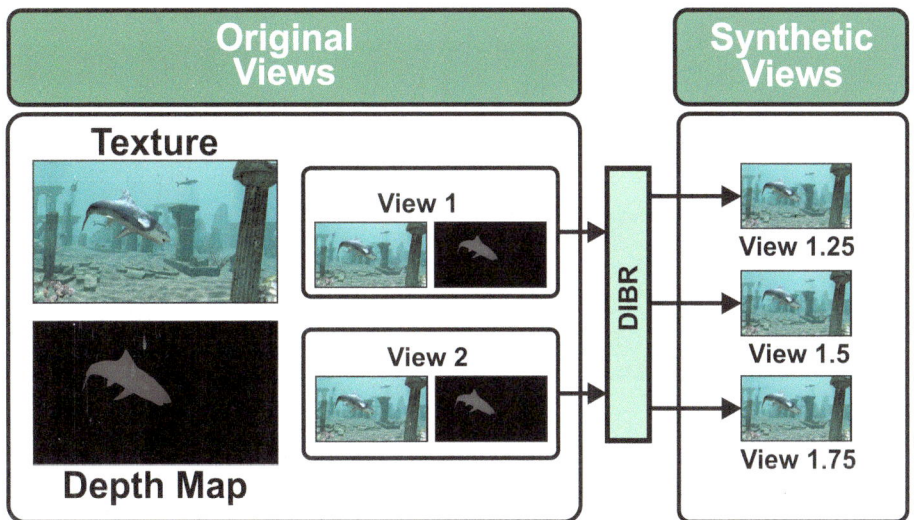

Fig. 1.1 Example of texture pictures and their respective depth maps considering a system with two original and three synthetic views

respective depth maps, considering a system with two original views, and the processing to generate three synthetic views from them using DIBR.

The pictures that compose the depth maps present well-defined characteristics, such as large homogeneous regions and sharp edges. This way, the encoding tools normally applied in 2D video encoding are not efficient enough to deal with depth-map encoding and improved tools are defined specifically to encode depth maps. The MVD approach also allows for exploiting the redundancies between different views through the technique called Disparity Estimation (DE). Therefore, the new coding tools introduced by the MVD approach to efficiently encode 3D-videos result in a higher computational effort when compared to the conventional 2D-video coding tools (Tech et al. 2016). Considering this expressive increase in computational effort, the support to handheld devices capable of dealing with MVD content tends to face severe constraints in terms of performance, storage, bandwidth and, mainly, energy consumption since these devices are battery-powered. Meeting real-time and energy constraints while sustaining high coding efficiency becomes more challenging when considering mobile battery-powered devices. While texture-only multiple view handheld camcorders are available in the market since 2010 (Panasonic 2024), mobile devices able to capture MVD content are expected to be widely available shortly. The popularization of handheld devices capable of dealing with 3D content is expected driven by emerging acquisition and displaying technologies like Epson Moverio BT-200 Smart Glasses (Epson 2024), Microsoft HoloLens 2 (Microsoft Hololens 2024), Intel RealSense 3D (Intel Realsense 2024), Structure Sensor 3 (Structure Sensor 2024),

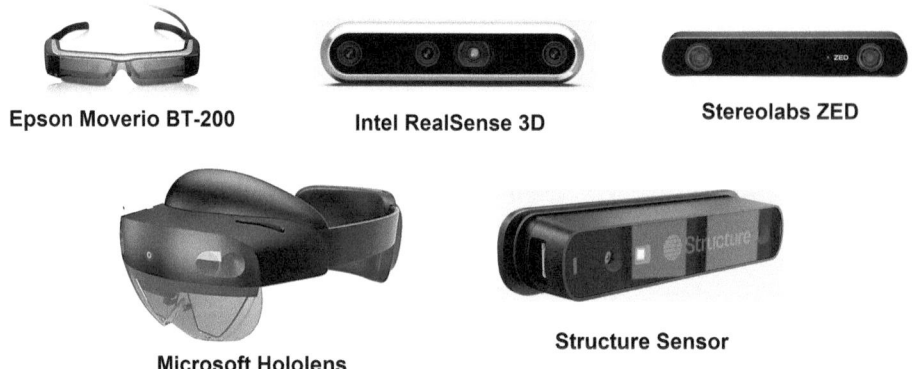

Epson Moverio BT-200   **Intel RealSense 3D**   **Stereolabs ZED**

Microsoft Hololens   **Structure Sensor**

Fig. 1.2 Examples of electronic devices based on the emerging acquisition and display technologies

and Stereolabs ZED 2 (Stereolabs ZED 2024). Figure 1.2 shows examples of electronic devices based on the emerging acquisition and display technologies.

Thereby, dedicated VLSI design focusing on MVD video coding is mandatory to provide high-throughput and energy-efficiency real-time encoding. The memory and computational effort required for the novel prediction tools evidence the need for the development of complexity-reduction strategies and VLSI designs. This book focuses on this research and development challenge, presenting energy-efficient and high-throughput hardware architectures for MVD video coding tools focusing on the 3D-HEVC standard as a case study.

1.1   3D Video-Coding Standards

Three-dimensional videos require much more data to be represented than two-dimensional videos since 3D videos are constructed using multiple views, commonly captured from multiple cameras, which represent the same scene from different viewpoints. Thus, a set of 2D videos are encoded and simultaneously transmitted resulting in a considerable increase in the computational effort of the 3D encoder and the bandwidth for 3D-video transmission, when compared to 2D videos. Therefore, new approaches to surpass this drawback and enable real-time 3D-video processing have been investigated in the last years.

The first intuitive solution to encode 3D-videos is a simulcast coding scheme (Tech et al. 2016), where the frames are predicted from frames belonging to the same view, i.e., each view is independently encoded as a 2D video. Although simulcast is less complex than other encoding models that exploit the redundancies between neighboring views, this coding scheme is much less efficient.

On the other hand, the multiple views of the same scene (multi-view) allow applications like stereoscopic and autostereoscopic displaying, Three-Dimensional Television (3D-TV), and Free-viewpoint Television (FTV) that may require a large set of views. Such applications become unfeasible with a simulcast approach regarding high resolutions because the 3D encoded video information tends to increase proportionally with the number of views. In this case, the final bitstream is the sum of all bitstreams of each view, which results in a huge amount of data to be transmitted/stored, mainly considering videos at high and ultra-high resolutions.

In order to efficiently exploit the coding process of three-dimensional videos the first standardized solution was the Multiview Video Coding (MVC) (ITU-T 2008) that was defined as an extension of the H.264/AVC (Advance Video Coding) (ITU-T 2017). The main idea of MVC was to exploit the redundancies among neighboring views reducing the encoded data. The MVC uses the H.264/AVC tools to encode these multiple videos and explore the redundancies between neighboring views in a new process called Inter-view prediction (ITU-T 2008). The main tool used in the Inter-view prediction is the Disparity Estimation (DE), which is based on the ME (Motion Estimation) used to encode two-dimensional videos (Sullivan et al. 2013). For each block of the current frame in a view, DE searches for a similar block in neighboring views (Sullivan et al. 2013). DE application increases the coding efficiency in the multi-view context, and this was the first tool included in video coding standards to explore the specificities of 3D information. According to Merkle et al. (2007), at least 20% of the image blocks that compose the frames are more efficiently encoded by using Inter-view prediction. When compared to the simulcast approach using H.264/AVC, the MVC can reduce the data needed for video representation by about 20–50% (Zatt et al. 2013). However, in MVC, the bit rate increases in a linear proportion to the number of views (Merkle et al. 2007). Since multi-view applications require many views to guarantee high quality in the viewing experience, the MVC should be improved to handle with this scenario.

The HEVC standard also defined an extension for multi-view tridimensional videos named MV-HEVC (Multi-view HEVC) (Tech et al. 2016). MV-HEVC follows a similar encoding structure of H.264/AVC MVC but using the new HEVC tools. Then, the MV-HEVC is also able to explore inter-view redundancies, increasing the coding efficiency. The MV-HEVC requires a higher computational effort when compared to MVC because the HEVC encoding tools are much more complex (and efficient) than the H.264/AVC encoding tools. However, the scenario with many views is also a problem that is difficult for the MV-HEVC to handle.

Aiming at solving this problem, experts of the JCT-3V (Joint Collaborative Team on 3D Video Coding Extension Development) developed a second HEVC extension targeting 3D videos, the 3D-HEVC (Tech et al. 2016). By using the Multi-view plus Depth (MVD) concept (Kauff et al. 2007), where a depth map is associated with each texture frame, the 3D-HEVC can be considered one of the current video coding standards developed to deal with 3D videos. The use of MVD allows reducing the number of captured and encoded texture views since the texture views, together with its depth maps, can be used at the decoder side to generate synthesized (or virtual) views, as previously explained.

Since the 3D-HEVC must consider the depth maps together with texture, novel encoding tools were also defined to efficiently deal with this type of information. These tools will be detailed in Chapter 2, including the new encoding modes for Intra-frame prediction: Depth Modeling Modes 1 and 4 (DMM-1 and DMM-4) and Depth Intra Skip (DIS); and an alternative residual flow Segment-wise DC coding (SDC) (Tech et al. 2016). All the original HEVC tools were also inherited and are available to be used in the 3D-HEVC.

Another important point considered in the development of the 3D-HEVC extension is the compatibility with the current technologies besides the technologies in development. In other words, 3D-HEVC is compliant with conventional 2D displays, stereoscopic displays, and autostereoscopic displays with a different number of views. The decoder can select a sub-bitstream according to the technology used in the display.

MPEG Immersive Video (MIV) standard emerged with the main goal of allowing the compression of a real or virtual 3-D scene representation, captured by multiple real or virtual cameras (Boyce et al. 2021). Therefore, MIV was developed focusing on immersive video coding and it also uses the MVD approach, such as the the 3D-HEVC.

An important characteristic of immersive video playback is the possibility of the viewer to control the view position and orientation of the content. In this context, other approaches as the 360◦ video brings limitations such as the supporting to three degrees of freedom, representing the orientation. Therefore, 360◦ video is not capable of supporting motion parallax, in which the relative position of objects changes based on the viewer's position with respect to the objects. On the other hand, immersive video provides playback with six degrees of freedom (6DoF) of view position and orientation within a limited range of motion, allowing the support to motion parallax, which improves the viewers' experience and avoids discomfort and even sickness for some viewers.

MIV supports the compression of immersive videos acquired by physical high resolution video camera systems, allowing a viewer's experience related to a real-world 3-D scene and many virtual reality (VR), augmented reality (AR), and mixed reality (MR) use cases. The MIV encoder processes multiple input views and enables the rendering of any intermediate views selected by the viewer. The major difference between MIV and MV-HEVC or 3-D HEVC codecs is that the MIV has been developed to require a few number of samples in order to not exceed the capability of devices. Also, The MIV coding framework was designed to accommodate any camera arrangement, with the encoder selecting the most appropriate information to be signaled. In other words, MIV was designed to avoid restricted viewing area, while reducing the necessary computational complexity targeting handheld devices, including simple smartphones with a viewport controlled by a touch screen and VR devices.

1.2   Contributions

This book presents dedicated algorithms and architectures for 3D video coding considering the multiview plus depth video content and focusing on the 3D-HEVC tools as a case
study.

The **major goal** was to improve the energy efficiency of the most energy-demanding
3D-HEVC encoding tools to make feasible the real-time processing of high and ultra-high
definition 3D-videos focusing on battery-powered applications. This goal was reached
through the exploration of memory/processing aspects, as well as the characteristics of
the encoding tools under an MVD approach.

The specific goals are listed below and they guide the insights and ideas exploited in
the proposed algorithms and in the designed architectures.

Goal 1: Identify the most time-demanding encoding tools according to the 3D-HEVC
for both texture and depth maps to assist the understanding of its energy requirements.
Goal 2: Identify the most selected and representative encoding tools during a 3D-
HEVC-based coding process to discover which tools have the most impact regarding
compression and image quality.
Goal 3: Evaluate the impact regarding compression by constraining specific encoding
tools as well as the block sizes supported by them during a 3D-HEVC-based coding
process.
Goal 4: Analyze, at run time, on-/off-chip memory access behaviors related to
3D-HEVC encoding process to define memory management and hardware design
methodologies capable of attempting the requirements regarding throughput and energy
consumption.
Goal 5: Take advantage of application-specific knowledge of 3D-HEVC (i.e., its new
coding tools) and video content properties to develop hardware-oriented algorithms
capable of increasing the throughput and improving the energy efficiency.
Goal 6: Use the developed hardware-oriented algorithms and the clock gating
technique to develop energy-efficient video memory and processing architectures.
Goal 7: Exploit a flexible coding order between texture and depth maps to propose heuristics and memory management capable of reducing the energy consumption
required for the developed energy-efficient video memory and processing architectures.

Figure 1.3 summarizes the main contributions of this book, including both algorithmic and
architectural levels. The proposal of algorithms and architectures was based on exhaustive
evaluations using the 3D-HEVC reference software, and this extensive evaluation itself is
the first contribution of this book.

As presented in Fig. 1.3, six architectures/systems exploiting memory/processing
aspects were proposed, four of them focusing on the Intra-frame prediction tools and
the other two focusing on the Inter prediction tools, the latter including both Inter-frames

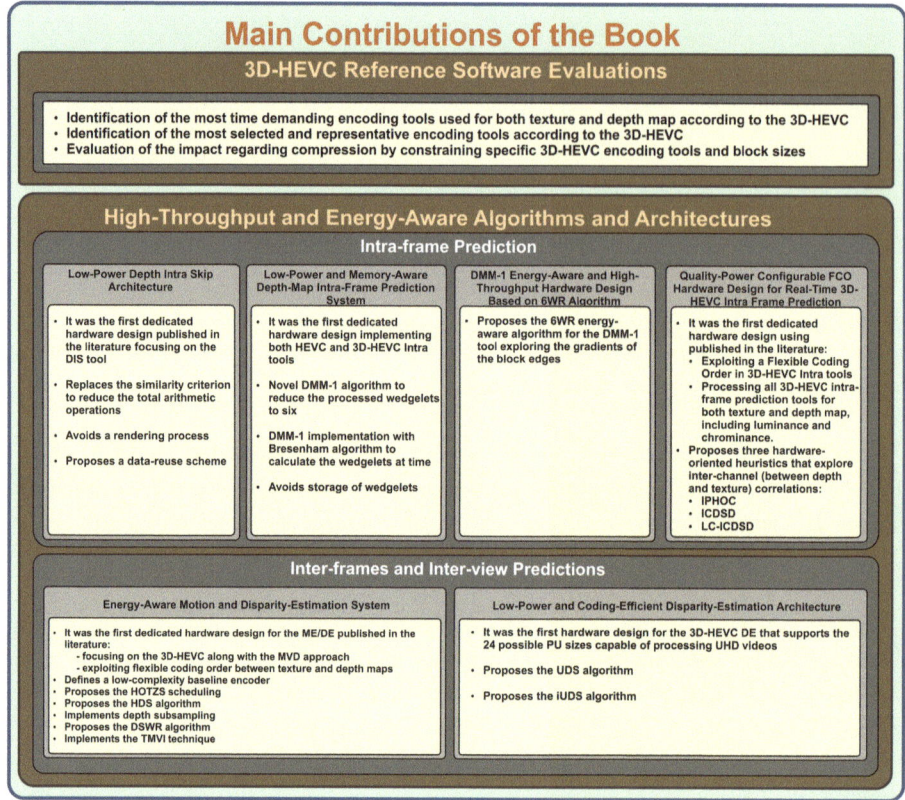

Fig. 1.3 Main contributions of the book

and Inter-view predictions. All the developed architectures take advantage of application-specific knowledge of 3D-HEVC along with the MVD approach, i.e., knowledge of its new coding tools and video content properties.

The first development contribution of this book is a low-power Depth Intra Skip (DIS) architecture that was the first dedicated hardware design published in the literature for the DIS tool (Afonso et al. 2017), and it is capable of processing five UHD 2160p views at 60 frames per second. This architecture used a simpler similarity criterion in order to reduce the computational effort. This strategy reduced the number of arithmetic operations and avoided a rendering process. Data-reuse schemes were also proposed in order to reduce energy consumption. This architecture is detailed in Sect. 4.1.

The second development contribution presented in this book is a low-power and memory-aware depth-map Intra-frame prediction system, which is also the first published solution supporting both the novel 3D-HEVC and the conventional HEVC intra prediction tools (Ücker et al. 2020). This architecture is capable of processing nine HD 1080p views at 30 fps. Hardware-oriented heuristics that consist of removing the less important modes and block sizes were proposed to reduce the computational effort. Also, a specific algorithm was applied to the Depth Modeling Mode–1 (DMM-1) to reduce the computational effort related to this tool. Section 4.2 presents this architecture.

The third development contribution presented in this book is a high-throughput and energy-aware hardware design for the 3D-HEVC Depth Modeling Mode 1 (DMM-1) based on the proposed 6WR (Six Wedgelets and six Refinements) algorithm (Perleberg et al. 2020b). The hardware design implements the Bresenham algorithm to avoid the use of memory. This hardware design is capable of processing up to nine views in 3D full HD 1080p videos at 30 frames per second. This architecture is detailed in Sect. 4.3.

The fourth development contribution presented in this book is a complete 3D HEVC intra-frame prediction hardware design that supports a flexible coding order between texture and depth channels (Perleberg et al. 2022). The developed hardware employs hardware friendly constraints and novel heuristics to explore inter channel redundancies and to reduce the computational effort through the novel Inter-channel Directional Structure Detector heuristic. The designed 3D-HEVC intra frame prediction system can process three HD 1080p views (texture + depth) at 30 frames per second in real-time. Section 4.4 presents this architecture.

The fifth development contribution presented in this book is an energy-aware Motion and Disparity Estimation (ME/DE) system that was the first hardware design focusing on the 3D-HEVC (with the MVD approach) published in the literature (Afonso et al. 2019a). This design is capable of processing three HD 1080p views at 30 frames per second. Also, this system was the first to take advantage of a flexible coding order between texture and depth maps to reduce energy consumption related to the 3D-HEVC ME/DE encoding tools. The architecture was designed for low-energy consumption, featuring a run-time adaptive memory hierarchy. Several heuristics and memory management capable of reducing the energy consumption were proposed. These and the other proposed techniques are described with more details in Sect. 5.1.

The last development contribution presented in this book is a low-power and coding-efficient Disparity Estimation (DE) architecture, which was the first published hardware design focusing on 3D-HEVC that supports 24 possible PU sizes capable of processing five UHD 2160p views at 40 frames per second (Perleberg et al. 2020a). Two low-complexity algorithms for the disparity estimation were proposed, prioritizing horizontal search instead of using conventional search algorithms in two dimensions. Section 5.2 presents this architecture.

1.3 Outline

This book is organized into six chapters, as follows:

Chapter 2 presents a 3D-HEVC overview. First, the 3D-HEVC basic coding structure is discussed. In the sequence, an HEVC encoding tools revision is shown. After, the novel 3D-HEVC encoding tools are presented for dependent-view encoding, depth-map encoding and the encoder control. Finally, this chapter presents the main ideas of state-of-the-art works regarding the development of algorithms and dedicated hardware architectures focusing on 3D-video coding. The works focusing on the development of energy-efficient systems and focusing on the 3D-HEVC are prioritized for such discussion. Promising works focusing on previous 2D/3D video coding standards are also presented.

Chapter 3 shows detailed 3D-HEVC reference software evaluations to motivate the algorithms and dedicated architectures available in this book. The main goal is to have a better understanding of the 3D-HEVC encoding-tools behavior to support the ideas developed in this book.

Chapter 4 presents the four dedicated architectures designed for focusing on the 3D-HEVC Intra-frame prediction. The first one is a low-power and high-throughput hardware design for the DIS coding tool. The second architecture is a low-power and memory-aware depth-map Intra-frame prediction system. The third architecture is an energy-aware and high-throughput hardware design based on 6WR algorithm for the DMM-1 coding tool. The last architecture is a quality-power configurable flexible coding order hardware design for the 3D-HEVC Intra-frame prediction.

Chapter 5 presents the two dedicated architectures designed to focus on the 3D-HEVC Inter-frames and Inter-view predictions. The first architecture is a ME/DE system designed for low-energy consumption, featuring a run-time adaptive memory hierarchy using the developed DSWR (Depth-Based Dynamic Search Window Resizing) algorithm. The second one is a low-power and coding-efficient Disparity Estimation architecture based on the developed iUDS (Improved Unidirectional Disparity-Search) algorithm.

Chapter 6 concludes the book by presenting the final remarks. All contributions are summarized, and the initially defined goals are discussed, based on the achieved results. As a reflexive analysis of the book, future research perspectives are presented.

3D-HEVC Overview

In March 2011, a call for proposals (ISO/IEC 2011) was disclosed in parallel to the HEVC standardization aiming to develop a 3D video-coding standard capable of working with an arbitrary number of views, when maintaining high efficiency in compression and image quality. Among several proposals, one proposal based on the HEVC standard and the concept of depth maps was chosen to be the initial point for the development of a new 3D video-coding standard.

After the choice of the baseline proposal, the group JCT-3V was created in 2012 from the collaboration between the experts of the VCEG (*Video Coding Experts Group*) of ITU-T (*International Telecommunication Union—Telecommunication Standardization Sector*) and the MPEG (*Moving Picture Experts Group*) of ISO/IEC (*International Organization for Standardization/International Electrotechnical Commission*) having as goal the development of a 3D video-coding extension in the top of the HEVC standard. Before, VCEG and MPEG groups had collaborated on the HEVC standardization through the JCT-VC (*Joint Collaborative Team on Video Coding*), which started its activities in 2010 and generated the first version of the HEVC standard in April 2013.

The development of this 3D extension for the HEVC standard was needed since the algorithms used in the 2D video-coding tools are inefficient in a 3D video-coding context when an arbitrary number of views is considered. Even the MVC extension of the H.264/AVC, developed focusing on the coding of multi-view videos, presented non satisfactory results for the 3D technology advance, mainly in the function of the current demand for high and ultra-high-resolution videos and the rise of multi-view auto-stereoscopic displays (Müller et al. 2013). The first version of the HEVC standard containing the 3D extension was published in February 2015 (ITU-T 2015).

V. Afonso et al., *Hardware Design for 3D Video Coding*, Synthesis Lectures on Engineering, Science, and Technology, https://doi.org/10.1007/978-3-031-80232-4_2

In turn, 3D-HEVC introduces the concept of depth maps for the coding of 3D videos along with new video-coding tools in order to reach high efficiency regarding compression. The use of depth maps allows that, from two or three views (composed of texture pictures plus their respective depth maps), several synthetic views representing intermediate positions can be generated maintaining image quality in this processing (Müller et al. 2013).

The following explanations on the 3D-HEVC extension and its coding tools are in agreement with the documents provided by the JCT-3V: the document *"Test Model 11 of 3D-HEVC and MV-HEVC"* (Chen et al. 2015), and the 3D-HEVC standard document (Tech et al. 2015). This chapter presents in detail the extension 3D-HEVC. Initially, the coding structure is explained and, in the sequence, the encoding tools are presented. Finally, the related works are presented and discussed.

2.1   3D-HEVC Background

2.1.1   3D-HEVC Basic Coding Structure

2.1.1.1   Block-Partition Structure

A digital video consists of a sequence of independent pictures, captured with a specific time interval. These pictures are commonly called frames, and frames are composed of pixels. Pixels are physical points that compose the image, generally using three samples that correspond to the brightness and color information. These samples can change according to the system used to represent the colors, i.e., according to the Color Space (Ohm 2015). The frames can also be divided into blocks by the video encoders. These blocks can present different sizes according to the video-coding standard and, for the current standards, the block size varies during the coding process in order to improve the efficiency of the encoding tools. Figure 2.1 presents a sequence of temporal-neighboring frames where one of them is divided into blocks. The representation of a scene with 2D digital videos is based on two types of sampling: (i) spatial and (ii) temporal. The spatial sampling uses a matrix of pixels defined as video resolution, which is responsible for delimiting the frame size. Considering that the samples of two different videos, recorded under the same conditions, have the same bit wide, the quality of the video is better the bigger the video resolution since a higher number of pixels is used to represent the same image. Temporal sampling is related to the time interval used to capture the images that compose the video sequence. The higher the rate used to capture the images, i.e., the higher the temporal sampling, the better the perception of movement. However, this improvement in the perception of movement reaches a limit according to the capacity of the human visual system. In general, the literature indicates a minimal rate from 24 to 30 frames per second in image capturing to guarantee the sensation of movement to the spectators (Gonzales 2003). However, high and ultra-high resolutions, such as UHD 2160p,

Fig. 2.1 Sequence of temporal neighboring frames and a frame divided into blocks

require a minimum of 50 to 60 frames per second to allow proper viewing (Goldman et al. 2015).

One of the main innovations presented by the HEVC when compared to the H.264/AVC is in the structure of compression. The adopted structure uses a highly flexible scheme of representation with three different concepts of blocks: (i) Coding Units (CUs); (ii) Prediction Units (PUs); and (iii) Transform Units (TUs; McCann et al. 2014). This way, each coding tool is employed over a different block structure, allowing more efficiency in the coding process according to the tool.

Video coding standards older than the H.264/AVC used a fixed block size. The H.264/AVC introduces the concept of block partition by adopting a macroblock of 16×16 samples that can be divided into smaller blocks until the minimum size of 4×4 samples. In turn, HEVC enables a maximum block size of 64×64 samples, while the minimum block size was maintained with 4×4 samples. Therefore, the HEVC allows a range of block sizes greater than the H.264/AVC in the predictions, and transforms/quantization steps, as will be better explained in the following.

The coding structure used in the HEVC initially allows the partition of the frames into square-shaped blocks called Coding Tree Units (CTUs; McCann et al. 2014). The CTU size can be configured. However, the maximum block size used in the coding process must be smaller or equal to 64×64 samples. The CTU is composed of one luminance block along with two chrominance blocks. The concept of CTU is analogous to the concept of a macroblock in H.264/AVC.

So, on each CTU, the highly flexible scheme of representation previously mentioned is implemented. Each CTU is composed of one or more CUs. CUs always are square-shaped blocks with size $2N \times 2N$, where N can assume the values 4, 8, 16, and 32. Each CU can be recursively divided into four blocks of the same size considering up to four CU-depth

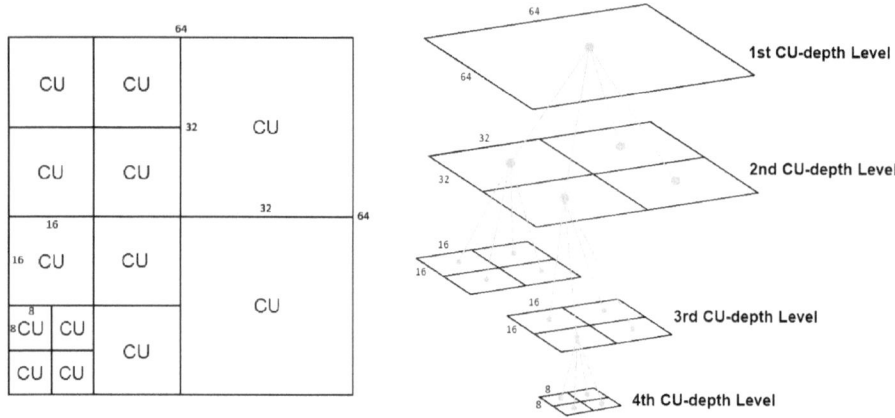

Fig. 2.2 Example of a CTU divided into CUs and its respective quaternary tree

levels from the CTU size. Hence, these divisions form a quaternary tree coding structure (quadtree). Figure 2.2 shows a CTU divided into CUs and its respective quaternary tree.

Each CU can have one or more PUs. The PU is the base block unit used for the Inter-frames and Intra-frame predictions. Unlike the CUs, PUs allow rectangular-shaped blocks. Hence, the blocks can be partitioned according to the real limits of the objects that compose the image. Figure 2.3 shows all the different ways that a CU can be divided into PUs. The values of N can be the same values previously presented for the CU sizes (4, 8, 16, and 32). The division N × N is only enabled in the Intra-frame prediction and when N is equal to 4. In the Intra-frame prediction, only N × N and 2N × 2N partitions can be used, while the Inter-frames prediction can also use symmetric partitions (2N × N, and N × 2N) and asymmetric partitions (2N × nU, 2N × nD, nL × 2N and nR × 2N). Asymmetric partitions are available only if N is bigger than four.

The TU is the base block unit used for the processes of Transform and Quantization. TUs are always square-shaped blocks of size N × N, where N can have the same sizes applied to the CUs (4, 8, 16, and 32). Each CU can have one or more TUs so that the TUs can also be disposed of a quaternary tree structure such as the CUs.

As mentioned above, the HEVC allows the partition of frames into blocks. However, the HEVC also allows the division of the video sequence into other hierarchy levels. In HEVC, Group of Pictures (GoP) are composed of frames, frames are divided into CTUs, CTUs are partitioned into CUs and, finally, the CUs are partitioned into PUs and TUs intending to reach the best coding efficiency. One GoP uses as references only frames inside the same GoP.

Figure 2.4 illustrates a video sequence divided into GoPs of size four. The concepts of I, P and B-frames are presented in the next subsection.

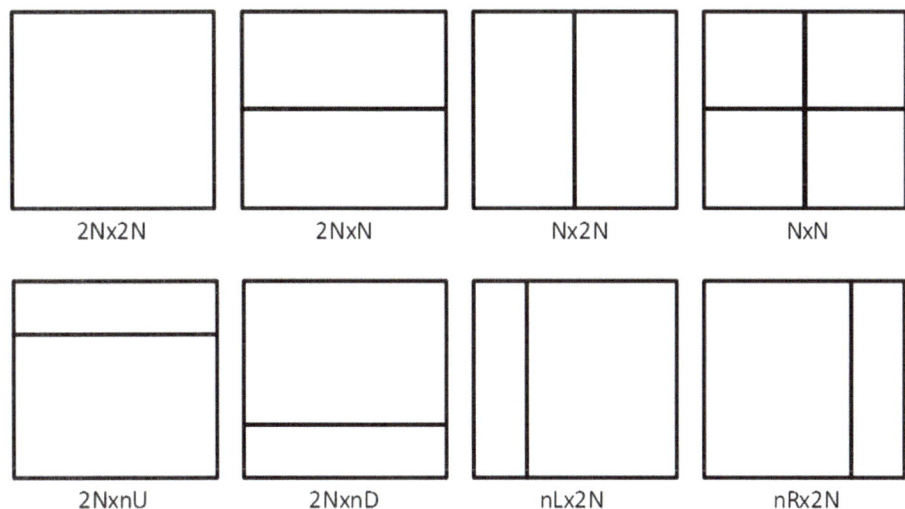

Fig. 2.3 Types of partition of a CU in PUs (McCann et al. 2014, p. 14)

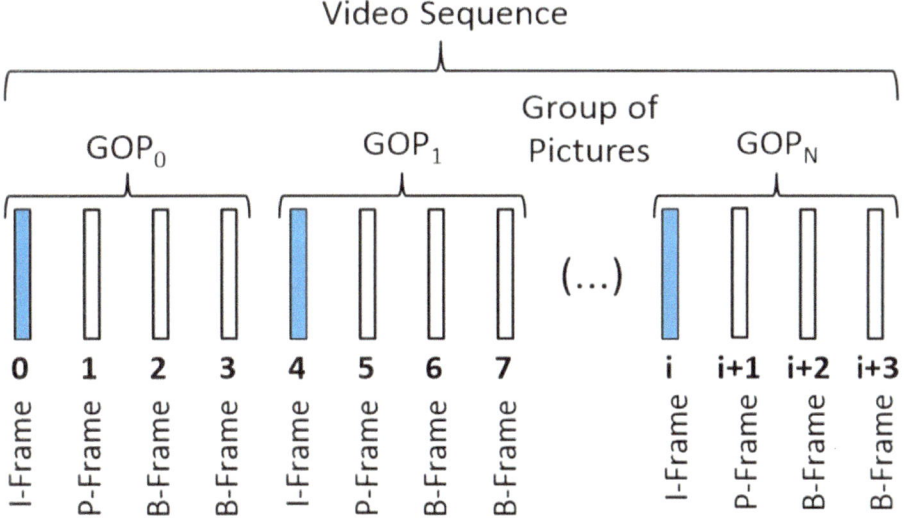

Fig. 2.4 Example of a video sequence divided into GoPs

2.1.1.2   Temporal Prediction Structure

The HEVC Reference Software (HM—HEVC Test Model) can operate under three different temporal prediction structures according to the experimental conditions, as defined in
the document JCT-VC L1100 (Bossen 2013). These structures are called intra-only (IO or

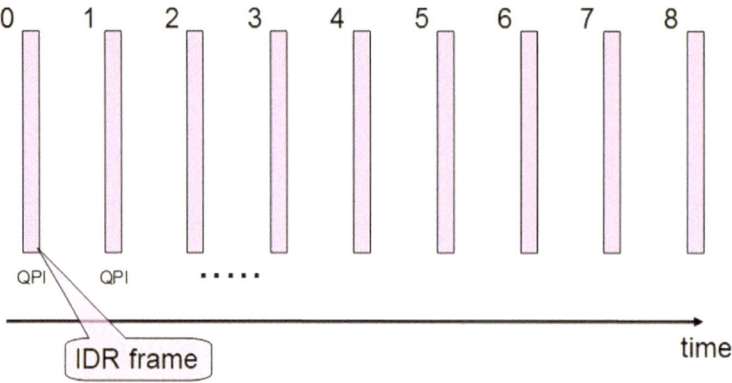

Fig. 2.5 IO temporal configuration (McCann et al. 2014)

All Intra depending on the JCT-VC document), low-delay (LD) and random-access (RA). HEVC extensions, such as the 3D-HEVC, can only operate under RA and IO temporal structures, then, only these two structures will be detailed in this text.

When operating under the test configuration IO, each frame that composes the video sequence is encoded as an IDR (Instantaneous Decoding Refresh). IDRs are frames predicted using only the Intra-frame tool. This way, no reference frames are used in the IO temporal structure. The QP value remains the same during the encoding process. Figure 2.5 represents the IO configuration, where the number associated with each frame represents the encoding order.

Considering the RA configuration, a hierarchy structure composed of bi-predicted frames (B frames) is used in the encoding process. B frames use two reference frames together to generate the prediction. One IDR frame is inserted every second approximately. The frames located between two IDR frames in the viewing order are encoded as B frames. The Generalized P and B pictures (GPB) are used in the lowest level of the temporal layer. GPB frames use only temporally previous frames in relation to the current frame. The second and the third temporal layers consist of referenced B frames whereas the highest level of the temporal layer contains non-referenced B frames. The QP value related to each Inter-coded frame is obtained from an off-set summed to the QP of the Intra frame. This off-set depends on the temporal layer. The viewing order in the RA configuration is not the same as used in the encoding process. Figure 2.6 shows the RA configurations and the number associated with each frame represents the coding order.

2.1.1.3 Quality Comparison Metrics Used Inside Video Encoders

Video encoders can exploit several techniques in order to obtain higher coding efficiency. These techniques use metrics to compare the original video and the compressed video and, therefore, to define which coding mode must be used for each block being encoded. The

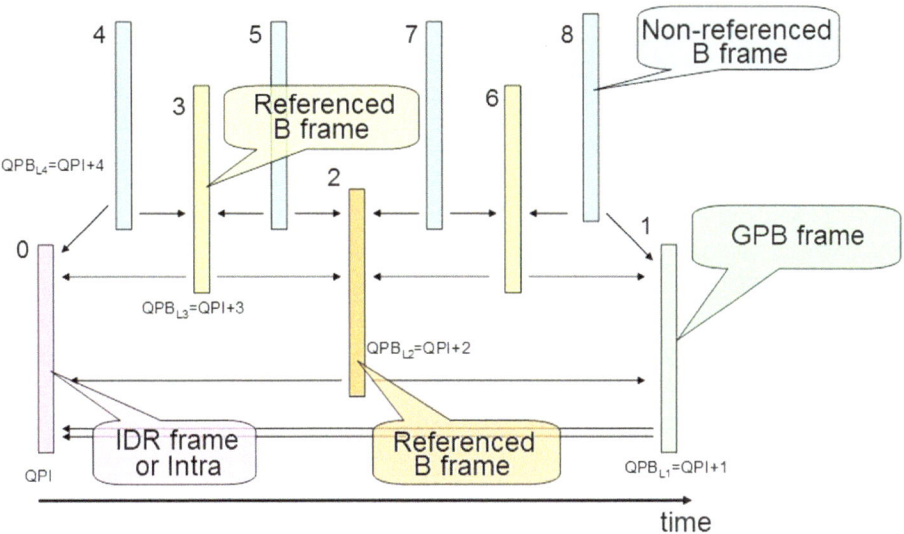

Fig. 2.6 RA temporal configuration (McCann et al. 2014)

coding decision is based on the relation between the reduction of data to be transmitted/stored (measuring the bit rate), and the impact of this compression on the image quality.

Current video-coding standards, such as the HEVC, use the Rate Distortion Optimization (RDO) technique (3D-HEVC Reference Software 2024) to measure the cost of the different encoding modes when applied to the candidate blocks in the encoding process, and decide which one returns the smallest cost. To obtain the distortion, RDO can apply different methods between the original and the reconstructed sample values, as will be explained in the following.

The objective criterion PSNR (Peak Signal-to-Noise Ratio; Ghanbari 2003), defined in Eq. (2.1), is the most used criterion in current video encoders (3D-HEVC Reference Software 2024). The PSNR can be applied in different levels from a block or frame to a complete video and it is represented in decibels (dB) that indicate the image-quality evaluated in the coding process. The PSNR calculation uses the similarity criterion called MSE (Mean Squared Error), as defined in Eq. (2.2). The MSE represents the mean squared error among the pixels of two frames. R and O represent the samples of the reconstructed and the original frames, respectively. The variables m and n determine the frame dimension, and MAX represents the maximum value of the samples. The higher is the PSNR, the better the objective image quality (Richardson 2002).

$$PSNR_{dB} = 20 * log_{10}\left(\frac{MAX}{\sqrt{MSE}}\right) \tag{2.1}$$

$$MSE = \frac{1}{mn}\sum_{i=0}^{m-1}\sum_{j=0}^{n-1}(R_{i,j} - O_{i,j})^2 \tag{2.2}$$

Besides the MSE, there are several other similarity criteria available in the literature. Among these criteria, the Sum of Absolute Differences (SAD; Seidel et al. 2016) is one of the most important in current video encoders (3D-HEVC Reference Software 2024). The SAD is a similarity criterion widely known in this community and it is commonly used in the prediction steps of many codec implementations. One advantage of SAD over MSE is that SAD allows an easier hardware implementation than the MSE, where only sums and subtractions are needed. The MSE requires costly processes like division and exponentiation operations. Equation (2.3) defines the SAD calculation, where R and O represent the samples from the reconstructed and the original frames, respectively. The variables m and n represent the frame dimensions.

$$SAD = \sum_{i=0}^{m-1}\sum_{j=0}^{n-1}|R_{i,j} - O_{i,j}| \tag{2.3}$$

Current video encoders (3D-HEVC Reference Software 2024) also employ a modi-fied version of the SAD criterion, the Sum of Absolute Transformed Differences (SATD; Seidel et al. 2016) that uses a Hadamard transform and imposes a higher computational effort than the SAD criterion. In other words, the SATD criterion delivers better results than the SAD, but the SATD also uses four times more arithmetic operations than the SAD (Grellert 2014). Equation (2.4) defines the SATD calculation where HT(i,j) repre-sents the application of the Hadamard Transform to the residual block obtained from the reconstructed samples and the original samples. The variables m and n represent the frame dimensions.

$$SATD = \sum_{i=0}^{m-1}\sum_{j=0}^{n-1}|HT_{i,j}| \tag{2.4}$$

2.1.1.4   3D-HEVC Coding Structure

Figure 2.7 presents an overview of a 3D-HEVC-based system that can be used to encode 3D videos, following an MVD format (Multi-view Video plus Depth). As presented in Fig. 2.7, 3D videos can be generated from different camera arrangements and signal sources containing either of the following: (i) a stereoscopic camera; (ii) a depth camera or a time-of-flight camera; (iii) an arrangement with n texture cameras; (iv) an arrangement with n depth cameras; (v) 3D content generated from a 2D-content conversion.

This way, before the pictures are coded with the 3D-video coder, they need to pass through a step of 3D-content generation. If the pictures have depth information (depth maps; from depth cameras, for example), they can be directly processed by the 3D

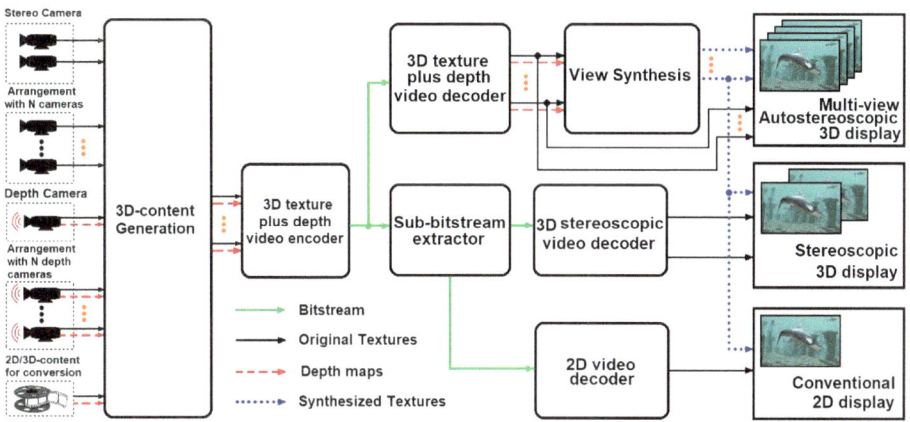

Fig. 2.7 Overview of the basic 3D coding structure in conformance with the 3D-HEVC extension

encoder. The depth maps associated with the texture pictures can be directly generated from cameras such as the time-of-flight (Li 2014a). However, if the depth map is not available, the texture pictures must pass through a depth-estimation step (Smolic et al. 2011).

Figure 2.7 also shows that the 3D-HEVC encoder generates a bitstream, which can be completely decoded to reconstruct the MVD video for a multi-view autostereoscopic display, but it also can be used to generate stereoscopic videos (only two views) or even 2D videos. This approach allows the reproduction of this content in different display technologies. After the decoding of the complete bitstream, the 3D decoder delivers a specific number of texture pictures, and their respective depth maps to the View Synthesis step. In turn, a stereo decoder can receive a sub-bitstream and deliver two channels of texture based on the original views to a stereoscopic display. Finally, a 2D decoder can provide 2D-content for a 2D conventional display from a sub-bitstream extracted from the bitstream transmitted by the 3D coder.

The View Synthesis step uses a rendering process called DIBR (Depth-Image-Based Rendering; Müller et al. 2013) to generate intermediate synthetic texture views between the original texture views. This way, a multi-view autostereoscopic display can be used to reproduce this content.

Figure 2.7 also shows that texture views can be provided for both 3D stereoscopic and 2D displays from the View Synthesis step.

The 3D-HEVC encoder organizes the input texture and depth map information using the concept of Access Units (AUs), as presented in Fig. 2.8. An AU contains all texture pictures and their respective depth maps belonging to the same time instant.

Each AU can have multiple views so that an index (*viewIdx*) is used to identify these views during the coding process. A number is attributed to each view index, i.e., the

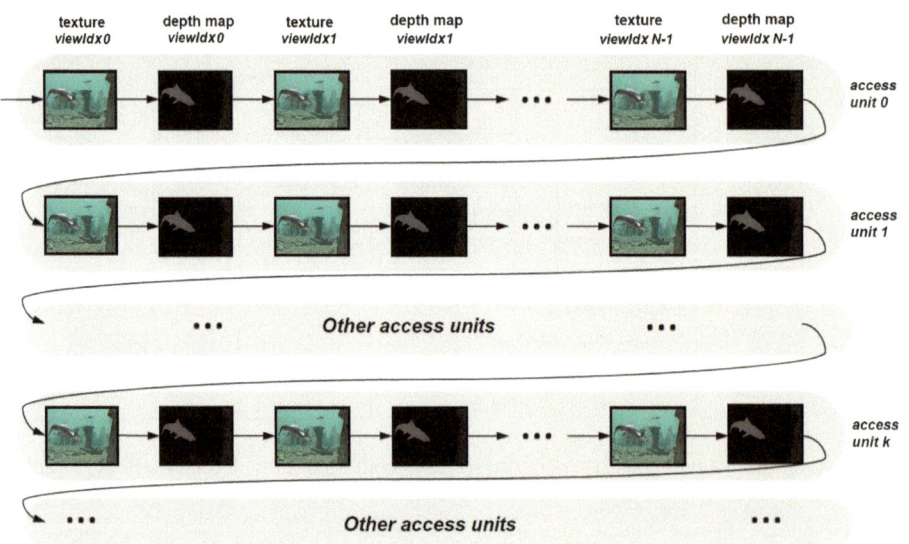

Fig. 2.8 Encoding process based on Access Units

texture picture and its respective depth-map receive the same number. The index can range from *0* to *n* − *1*, and the view that receives the index *0* is always the first one to be encoded in each AU. This first view to be encoded is called Base View (BV), or Independent View, and it is encoded without using information from other views. The other views belonging to an AU are called Dependent Views (DVs), and they receive index values between *1* and *n* − *1* according to the coding order. It is important to notice that in the BV, the texture picture is always encoded before its depth map. By default, texture pictures are also encoded before depth maps in DVs. However, this order can be inverted enabling the Flexible Coding Order (FCO; see Sect. 2.1.1.6). As presented in Fig. 2.8, each AU is also represented by an index number. The first AU receives the index *0*, whereas the following AUs receive indexes between *1* and *n* − *1*.

The 3D-HEVC inherits all block partition structures and the temporal prediction structure defined in the HEVC. In other words, an encoder based on the 3D-HEVC also divides the pictures (both textures and depth maps) into smaller square-shaped block sizes before applying the encoding tools. The same HEVC concepts are presented in 3D-HEVC, which covers the Coding Tree Units (CTUs), Coding Units (CUs), Prediction Units (PUs), Transform Units (TUs), and Group of Pictures (GoPs), among others.

Figure 2.9 shows a simplified 3D-HEVC Encoding Model in which some novel encoding tools and methods can be observed in addition to the encoding tools normally used in HEVC. Considering the Intra-frame prediction, the 3D-HEVC inherited all the HEVC prediction modes, i.e., Planar, DC and Angular modes, which are used to encode both texture and depth frames. The Intra-frame prediction exploits the spatial redundancies presented in a video. Considering the same frame of a video, most of the neighboring pixels tend to present similar values (or even similar behavior when the values smoothly change in a given direction). Besides, the 3D-HEVC Intra-frame prediction includes new tools to deal with depth maps, comprising the depth modeling modes one and four (DMM-1 and DMM-4) and the Depth Intra Skip (DIS).

In the Inter-frames and Inter-view predictions, Motion and Disparity Estimations are used to encode both texture pictures and depth maps using the same tools defined for the HEVC Inter-frames prediction. The Inter-frames prediction is responsible for exploiting the temporal redundancies. Temporal redundancies are those similarities that appear in neighboring frames in a video. These frames tend to have many similar or even identical regions. The Inter-frames prediction is composed of two main steps: the Motion

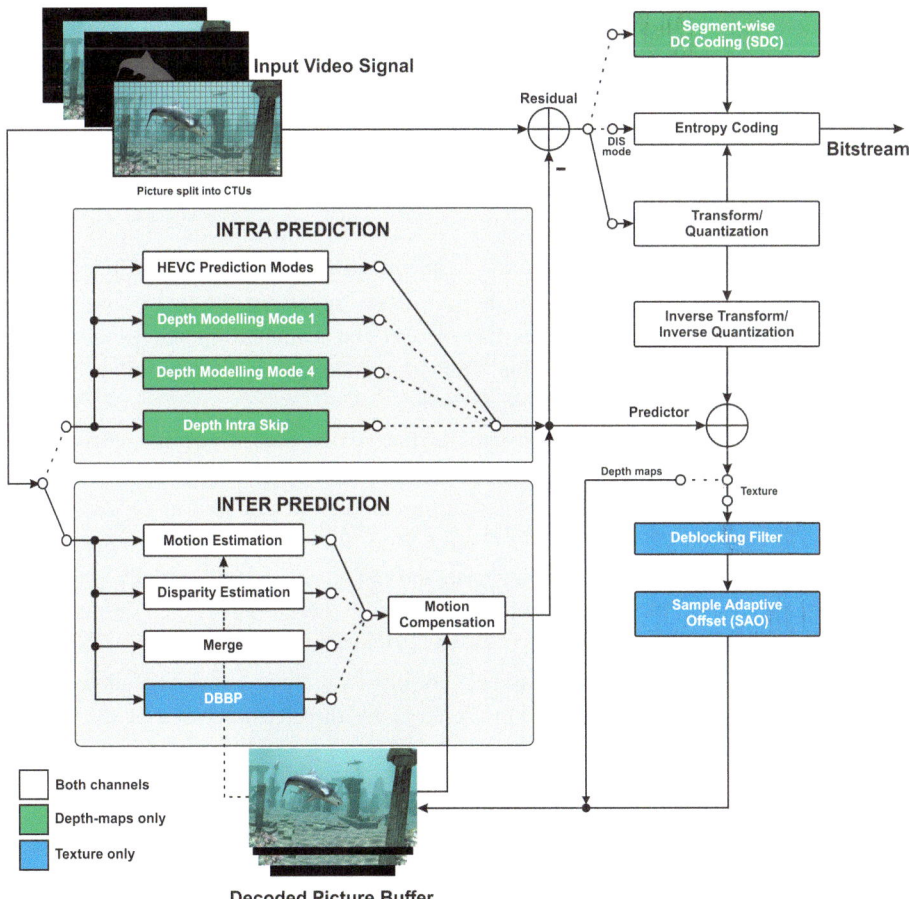

Fig. 2.9 Simplified 3D-HEVC encoding model

Estimation (ME) and the Motion Compensation (MC; Agostini 2007). ME is responsible for finding the block previously processed in the reference frames that is most similar with the block being encoded, considering some parameters to constrain the search to limit the amount of evaluated blocks and reduces the computational effort. MC is used to reconstruct the frames estimated from the data generated by the ME, allowing residue generation (the subtraction between the current block and the predicted block). By using the ME and MC, only the residues and the relative localization (motion vector) of the block selected in the reference frame are transmitted. The Disparity Estimation extends the ME concept to the neighboring views in a video. The Depth-Based Block Partitioning (DBBP) mode is applied only to the texture pictures.

A control step in the encoder is responsible for choosing the prediction tool that will be used to encode each block according to video characteristics, with the intention of maximizing the relation between compression and image quality. After the prediction, a subtraction operation is performed between each original image block and the respective predicted block to obtain the difference between these blocks, which is called residue.

In HEVC, the residue passes through a Transform step that transforms the information from the spatial domain to the frequency domain. This way, an additional step called Quantization eliminates those frequencies that are less relevant to the human visual system. This operation introduces losses in terms of image quality (Agostini 2007). Nevertheless, these losses can be controlled by the encoder optimizing the relation between the image quality and the compression rate. The Quantization Parameter (QP) defines the used quantization level, and the higher the QP value, the higher the image compression and the lower the image quality. Then, the encoder can define an operation point where the image degradations are imperceptible. Transform and Quantization operations together also contribute to reduce the spatial redundancies.

The 3D-HEVC residual encoding supports the HEVC Transform and Quantization steps, as well as the other two alternative methods. The first one is the Segment-wise DC Coding (SDC), which can be applied to all prediction modes used in the encoding of depth maps, except the DIS tool. The SDC method consists of a residual encoding method in which the residual data bypasses transform and quantization steps. The DC residue of a PU is calculated from the average of the differences between the original sample values and the predicted sample values. SDC can be applied only when a PU uses the 2N × 2N partition. The second one is applied only for blocks predicted with DIS mode, where the residual encoding is bypassed, and the residues are directly processed by the Entropy Coding.

The Entropy Encoding is applied to change the representation of the symbols using the variable-length code. Thus, the symbols with higher occurrences are represented with a smaller bit width. On the contrary, symbols with lower occurrences are represented with a higher bit width. Then, this tool is responsible for reducing entropic redundancy, which is the third redundancy presented in digital videos. This redundancy is directly related to the number of bits necessary to represent the video information. The HEVC main tool used in the Entropy Encoding is the CABAC (Context-Based Adaptive Binary Arithmetic Coding; McCann et al. 2014).

The encoder also has the Inverse Transform and Inverse Quantization steps that are used in the generation of the reconstructed image blocks. The reconstructed image blocks compose the reconstructed frame, which allows the next frames to be encoded and guarantees that both the encoder and the decoder use the same references since the quantization generates losses in the encoding process (Agostini 2007).

Finally, the In-loop filtering step is used to reduce the block effects introduced due to the block-based coding, which differentiates a real image edge from an artifact generated by the Quantization. Two filters are used in HEVC before the reconstruction of samples: (i) the Deblocking Filter and (ii) the SAO (Sample Adaptive Offset). These filters increase the subjective image quality (McCann et al. 2014).

2.1.1.5   Conventional 3D-HEVC-Based Encoding Process

Figure 2.10 shows a basic structure of the 3D-HEVC encoder where one can observe that the texture picture of the BV (*view 0*) is processed with a conventional HEVC encoder. This way, the decoder can provide pictures to a conventional 2D display. The other texture pictures and the depth maps are encoded using HEVC-based encoders, where additional coding tools are inserted to explore the inter views and inter components redundancies and, also the depth maps characteristics, as explained in the next paragraphs. The final bitstream is provided after a multiplexing step where the outputs of each encoder (one for each channel of each view) are switched for the system output.

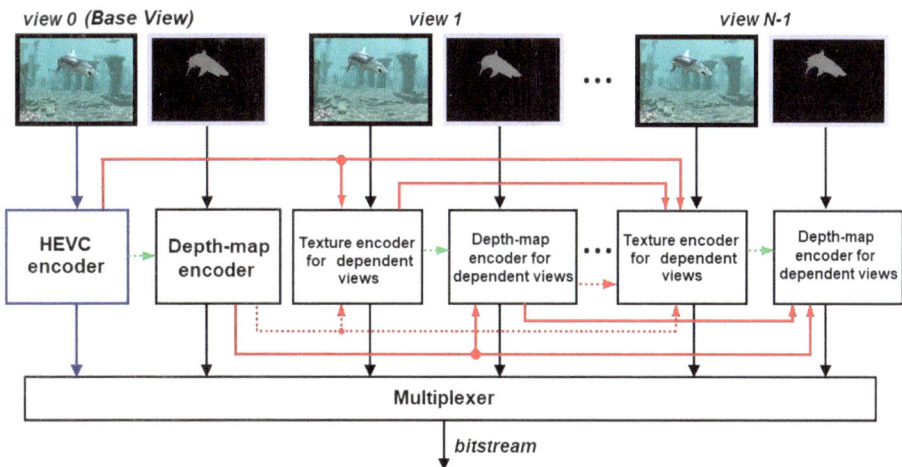

Fig. 2.10   CCO base structure of the encoder with inter-component and inter-view predictions

The Inter-component prediction is the tool responsible for exploiting the redundancies between different channels, i.e., between texture and depth-map channels. Texture information can be used to encode the depth maps in the same view, represented by the dotted green arrows in Fig. 2.10 (used in the DMM-4 mode, for example—see Sect. 2.1.4.1.2) and, also, the depth-map information can be used to encode the texture of other views, represented by the dotted red arrows in Fig. 2.10 (used in the DBBP mode, for example—see Sect. 2.1.3.3). However, the Inter-component prediction can be disabled and, in this case, the texture pictures are decoded without using the depth maps information. The Inter-view prediction explores the redundancies between different views but considers the same channel and it is represented by the continuous red arrows in Fig. 2.10. Basically, the Inter-view prediction is composed of the Disparity Estimation tool (see Sect. 2.1.3.1).

Figure 2.10 shows the Inter-component and Inter-view predictions considering the conventional 3D-HEVC encoding process, here called conventional coding order (CCO).

As previously mentioned, BVs are encoded through an encoder in conformance with the HEVC. In the case of DVs, the same concepts and encoding tools are used. However, data of views previously encoded are used to enhance the coding efficiency of DVs by adding new coding tools to the HEVC encoder. Therefore, Sects. 2.1.3 and 2.1.4 present these novel encoding tools in detail.

2.1.1.6 Encoding Using Flexible Coding Order (FCO)

The Flexible Coding Order (FCO) for the 3D-HEVC was proposed in the work (Gopalakrishna et al. 2013). FCO features the encoding of depth maps before their associated texture pictures for the dependent views, i.e., the first view (denoted Base View or Independent View) is encoded in a conventional way (texture before depth map) and in the remaining views, that used information from other views, the depth maps can be encoded first, as depicted in Fig. 2.11.

A conventional 3D-HEVC coding order (CCO) improves the compression efficiency of dependent views by using disparity information derived from the motion information of the neighboring blocks (additional candidates tested in the Merge mode). The idea brought by the FCO configuration is that disparity information can be derived directly from the depth data, increasing the accuracy of the disparity vectors. This method with a different encoding order is depicted by the dotted brown arrows in Fig. 2.11, where disparity information is used to aid the prediction of the current texture block. The work (Gopalakrishna et al. 2013) measured the impact of the FCO method regarding compression and it decreased by 0.6% the BD-Rate of video total, on average. In other words, this approach improves the compression performance of 3D-HEVC. Also, the proposed configuration reduced the complexity of decoding by over 8%.

In this book, the FCO configuration was exploited in the hardware designs presented in Sects. 4.1 and 5.1. In Sect. 4.4 the hardware mainly exploits the correlations between the Angular modes and the novel DMM modes. In Sect. 5.1 the goal was to obtain energy efficiency for the developed ME/DE system. This feature enables the depth information

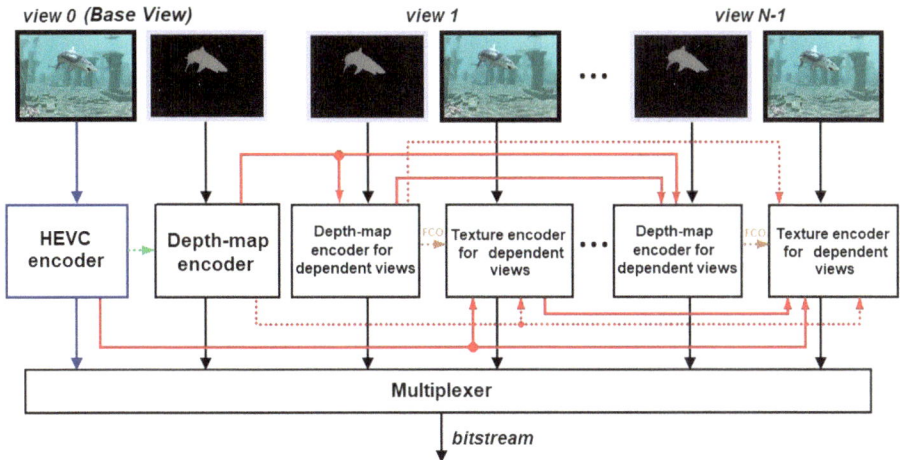

Fig. 2.11 FCO base structure of the encoder with inter-component and inter-view predictions

to manage the memory and processing units of the developed system according to the channel that is encoded first.

2.1.2 HEVC Encoding Tools

In terms of coding tools, HEVC introduces some novelties when compared to the H.264/AVC. The following subsections discuss these novelties, focusing on both Intra-frame and Inter-frames predictions, which are the focus of the contributions presented in this book.

2.1.2.1 Intra-Frame Prediction

In the Intra-frame prediction, the HEVC maintained the DC mode and the eight directional modes used in the H.264/AVC, but introduced the Planar mode and other 25 new directional modes. Therefore, the total number of modes increased from nine in the H.264/AVC to 35 in the HEVC. As previously explained, the Intra-frame prediction is responsible for exploiting the spatial redundancies presented in neighboring regions inside the same frame.

The next subsections explain the intra-prediction encoding tools.

Planar Mode (Mode 0)

The Planar prediction mode is efficient in predicting regions with smooth textures or complex textures that are not efficiently predicted by the angular modes (Lainema et al. 2014), and it results in predicted blocks with surfaces without discontinuities in the edges.

The Planar mode is calculated through the average of the results provided from the interpolation using two filters, called here: (i) Horizontal prediction (P$_h$), defined by Eq. (2.5); and (ii) Vertical prediction (P$_v$), defined by Eq. (2.6). These filters calculate their values from the multiplication of reference samples belonging to the adjacent blocks, located above and to the left of the block being encoded. Equation (2.7) is used to calculate the final value of the Planar mode for each position.

The value of N in Eqs. (2.5–2.7) represents the size of the block to be predicted, where x and y ϵ 0, ..., N − 1, while Lat[x] and Lat[N] are reference samples located to the left of the current block, Sup[y] and Sup[N] are samples located above to the current block and P$_h$[x][y] and P$_v$[x][y] are the samples generated by the filters.

$$Ph[x][y] = (N - 1 - x)*Lat[x] + (x + 1)*Sup[N] \qquad (2.5)$$

$$Pv[x][y] = (N - 1 - y)*Sup[y] + (y + 1)*Lat[N] \qquad (2.6)$$

$$P[x][y] = (Ph[x][y] + Pv[x][y] >> (log2(N) + 1) \qquad (2.7)$$

The Planar mode results are obtained by the average of horizontal and vertical filter results, as demonstrated in Eq. (2.7), where P[x][y] is the final result of the Planar mode.

Figure 2.12 shows the calculation of Horizontal and Vertical filters in an 8 × 8 block, as well as the use of the partial results for the Planar calculation.

DC Mode (Mode 1)

The DC prediction mode is capable of efficiently encoding homogeneous regions (Lainema et al. 2014). DC mode is based on the calculation of the value denoted *dcval* and, after that, this value is copied for all samples of the predicted block. If the block does not have any reference sample, the *dcval* value is changed for the middle value considering the bit-depth of the samples. That is, if the samples are eight bit, *dcval* is set to 128.

Angular Modes (Modes from 2 to 34)

The Angular modes were defined with the aim of efficiently modeling the directional structures presented in video textures (Lainema et al. 2014), where well-defined edges are presented (Sullivan et al. 2012). For that, angular modes generate the prediction blocks by projecting the neighboring samples in different directions.

The prediction with the Angular modes in HEVC is made similarly to its predecessor standard, the H.264/AVC, but with a larger number of reference samples, angles and block sizes. Each angular mode works with a precision of 1/32 (Lainema et al. 2014). The angles used in the HEVC standard were selected after evaluations, and therefore, they are not equidistant. The angles associated with angular modes can be seen in Fig. 2.13.

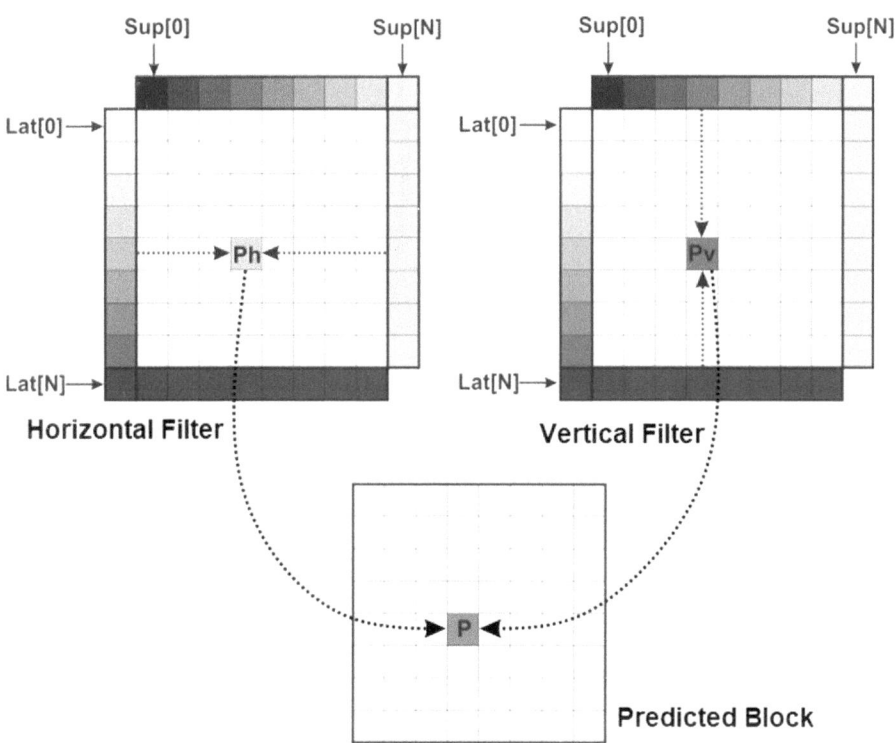

Fig. 2.12 Example of an 8 × 8 block calculated using the planar mode

The angular modes can be classified into vertical modes and horizontal modes. The horizontal modes (from 2 to 17) use mainly samples from the left of the block being predicted, whereas the vertical modes (from 18 to 34) use mainly samples above the block being predicted to represent the current block. To obtain the value of each sample of the predicted block, a set of equations is used to define which neighboring sample is used to compute the value of each sample from the predicted block, and also the multiplication factor when the weighted average of two samples is used.

Each angular mode has an associated pre-defined value called Angular Parameter (A) that varies from −32 up to 32 representing the direction of the prediction process. The neighboring samples of the current block are used to create a reference vector by simply copying the neighboring samples, based on Eqs. (2.8) and (2.9). However, the modes that have negative values for A (from 11 to 25) also have an associated pre-defined parameter B used to project some samples required for the prediction. In other words, the B parameter becomes necessary to select the samples that will be used to complete the reference vector in each mode, as presented in Fig. 2.14, where some specific neighboring samples are projected to complete the reference vector. The projection of the required samples to

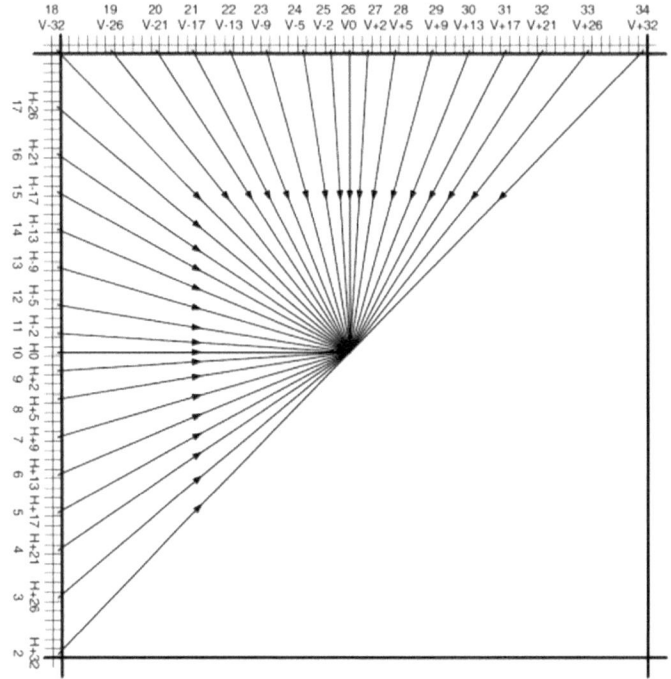

Fig. 2.13 Angular modes definition (McCann et al. 2014)

complete the reference vector can be computed by Eqs. (2.10) and (2.11), being x and y the indexes of the reference vector.

$$ref[x] = side[x], \text{ for horizontal modes} \tag{2.8}$$

$$ref[y] = top[y], \text{ for vertical modes} \tag{2.9}$$

$$ref[x] = top[((x * B) + 128) \gg 8], \text{ for horizontal modes} \tag{2.10}$$

$$ref[y] = side[((y * B) + 128) \gg 8], \text{ for vertical modes} \tag{2.11}$$

After generating the reference vector, the block is predicted by projecting the samples of the reference vector in the respective angle. In this projection, the value of the samples from the predicted block is computed by Eqs. (2.12) and (2.13), which are a weighted average of the two samples from the reference vector located in the direction of the respective angle.

Fig. 2.14 Projection of samples to complete the reference vector

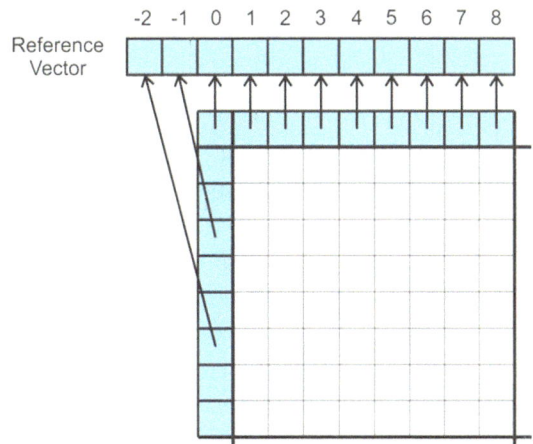

$$p[x][y] = \big((32 - f) * ref\big[y + i + 1\big] + f * ref\big[y + i + 2\big] + 16\big) \gg 5, \text{ for horizontal modes} \tag{2.12}$$

$$p[x][y] = \big((32 - f) * ref[x + i + 1] + f * ref[x + i + 2] + 16\big) \gg 5, \text{ for vertical modes} \tag{2.13}$$

In Eqs. (2.12) and (2.13), i represents a projection integer displacement on row y or column x, which is used to select different samples from the reference vector depending on which sample of the predicted block is being computed. The value of i can be computed by Eqs. (2.14) and (2.15).

$$i = ((x + 1) * A) \gg 5, \text{ for horizontal modes} \tag{2.14}$$

$$i = ((y + 1) * A) \gg 5, \text{ for vertical modes} \tag{2.15}$$

In addition, Eqs. (2.12) and (2.13) select two samples from the reference vector to compute each sample of the predicted block. These two samples are multiplied by an f value, which is the weight of each of the two selected neighboring samples, based on the distance from the predicted sample to the neighboring samples, represented by y or x. The value of f can be obtained by the Eqs. (2.16) and (2.17).

$$f = ((x + 1) * A)\&31, \text{ for horizontal modes} \tag{2.16}$$

$$f = ((y + 1) * A)\&31, \text{ for vertical modes} \tag{2.17}$$

Based on Eqs. (2.8–2.17), the current block can be predicted using the neighboring samples of the current block, based on any of the 33 angular modes used in the HEVC standard.

Despite the HEVC standard defining 33 angular prediction modes, the Mode 10 (Horizontal) and the Mode 26 (Vertical) have a particularity since they only perform a simple copy of the reference samples in the horizontal direction, for the Mode 10, and in the vertical direction, for the Mode 26.

2.1.2.2 Inter-Frames Prediction

As previously mentioned, HEVC employs a block-based hybrid video coding scheme. Each block is evaluated by several coding tools, and the most efficient one is selected. Among these tools, the Inter-frames prediction, with the ME and Merge modes, takes a prominent position in current video encoders due to the large compression gains achieved and the high energy consumption related to their intense processing and memory communication (as discussed in Sect. 3.1). Both ME and Merge are responsible for exploring temporal redundancies by performing block matching between frames at different time instants, as will be better explained in the sequence.

Motion Estimation (ME)

Motion Estimation (ME) stands as the most computational/energy/memory-intensive task within the encoder (Zatt et al. 2011b). Its goal is to exploit temporal redundancy by searching for similar blocks in previously encoded frames.

The ME consists of comparing the current block with reference blocks of previously processed frames in order to find the most similar one under a similarity criterion. The search pattern is determined by a BMA (Block Matching Algorithm) and it is typically constrained within a Search Window (SW). The TZS (Test Zone Search) algorithm (Jia et al. 2013) has become the most widely used BMA in the latest encoder implementations, whereas the SAD (Sum of Absolute Differences) is the most commonly used similarity criterion when real-time systems are targeted. The ME process can be observed in Fig. 2.15.

Additionally, to improve coding efficiency, HEVC defines the use of different block sizes (known as PUs) in the ME, ranging from 64×64 down to 4×8 pixels. Considering the full Rate-Distortion Optimization (RDO) process (3D-HEVC Reference Software 2024), the decision on which PU size must be used occurs after the complete evaluation of all possible combinations considering the 24 PU sizes, so that the best tradeoff between compression and distortion is achieved. Note that even when only 32×32 blocks are considered, the BMA should evaluate 90 million candidate blocks per second in order to allow real-time processing of HD (High Definition) 1080p 3D videos with three views (Afonso et al. 2019a). In this scenario, each evaluated candidate requires the fetching of 32×32 8-bit samples from the reference memory, leading to memory traffic of 92 GB/s.

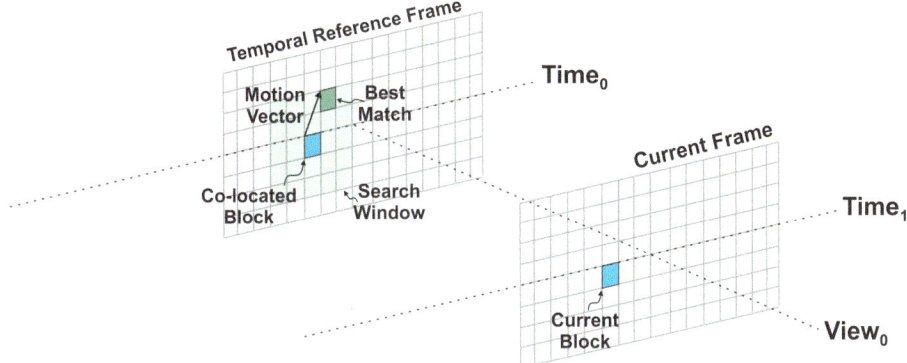

Fig. 2.15 Illustration of the ME process

Memory access emerges as one of the main bottlenecks regarding energy consumption and processing rate when considering an ME hardware design. On the one hand, a limited memory bandwidth harms the system performance due to the existence of many candidates to be compared by the BMA. On the other hand, huge memory bandwidth results in a significant power dissipation/energy consumption of the system bus (Chen et al. 2006). Therefore, data reuse must be adopted to reduce memory bandwidth while leaving the system performance unhindered.

The Level-C scheme allows the reuse of overlapping memory regions between neighboring SWs (Chen et al. 2006). It considers that, during the coding process, neighboring blocks share much of the SW. Thus, this data may be kept in the on-chip memory, avoiding external memory retransmission. Figure 2.16 depicts this process. When the encoder requests the Block-2 SW, the Exclusive Block-1 Region is discarded from on-chip memory, the Exclusive Block-2 Region is brought from the external memory, and the Overlapped Region is reused.

Note that although Level-C was proposed a few years ago, it is still widely used in dedicated ME solutions due to its great tradeoff between efficiency and ease of implementation. Furthermore, current related works are still proposing and implementing data-reuse schemes for on-chip memories based on Level-C (Jia et al. 2015) for their ME systems. Other works adopt Level-C as the baseline for comparison (Zatt et al. 2011a; Sampaio et al. 2013). Therefore, this data-reuse scheme was adopted as the baseline implementation, which allows for demonstration of the gains of the developed solutions for memory hierarchy and management implemented in this book, and also comparisons with the main related works.

As explained before, ME applies the BMA to select a set of reference blocks to be compared with the current block, and to find the most similar reference block to the current block. The output of the ME step is the information about the motion vector related to the best matching block, i.e., the most similar processed reference block to the

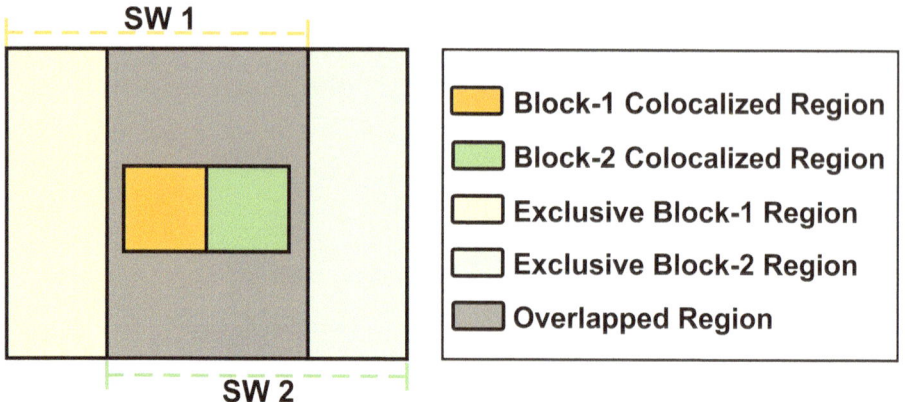

Fig. 2.16 Level-C scheme overview (Adapted from Afonso et al. (2019a))

current block. This information is used to reduce the data transmission/storage, since only the difference between the current block and the most similar reference block, and the motion vector related to this block, are necessary to represent the block being encoded.

To reduce the complexity of the ME, the BMA is applied only to a pre-defined and reduced search window, whose size depends on the CTU size and also on the search range (SR). There are several BMA in the literature with a focus on the ME step (Fan et al. 2018; Perleberg et al. 2018), each one with its own particularities. The HEVC Reference Software, the HM, adopts the TZS algorithm to perform the ME step. TZS is classified as a fast algorithm (Li et al. 2014b) because it processes a small set of reference blocks of a given SW, while it maintains the visual distortion practically unchanged when compared to the Full Search (FS) algorithm (Li et al. 2014b), which generates the optimal result for a given SW.

The TZS algorithm is composed of four steps, where each step result defines the start position of the next step. These steps are the Prediction, First Search, Raster and Refinement (Li et al. 2014b), and the flowchart of TZS steps is present in Fig. 2.17a. The Prediction is the first step of the TZS. It is composed of five different predictors that evaluate specific blocks based on their probability of being the best matching block. After this, the First Search is applied.

Both First Search and Refinement steps perform an expansive search which compares several blocks around their start point. The scheme of the expansive search is presented in Fig. 2.17b, where the colored circles represent the first sample of each candidate block. It starts evaluating the four blocks around the position of the start point (highlighted in yellow). Then, the search expands in a diamond scheme (the scheme can vary according to the default configurations of the HM) to evaluate candidates farther from the center until it reaches a stop condition.

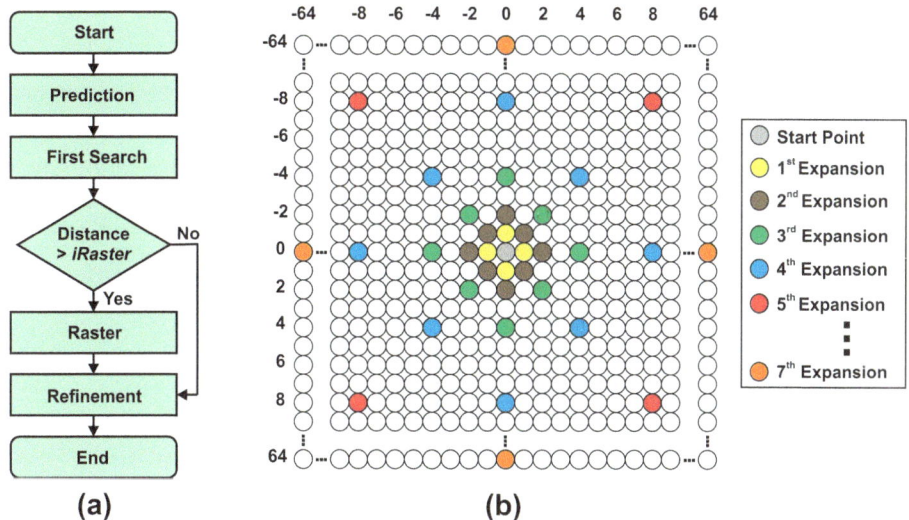

Fig. 2.17 TZS algorithm: **a** Flowchart; **b** Expansions

The First Search and Refinement step has two stop conditions that occur when the SW limit is reached, or when the algorithm expands three consecutive times without finding a better matching block. The First Search advances to the next step when any of these conditions occur. However, the Refinement only ends the iteration (set of expansions around a given position) when the SW limit is reached, or when the algorithm expands three consecutive times without finding a better matching block. In other words, the Refinement starts a new iteration around the best result found in the previous iteration when those stop conditions occur. This characteristic of the Refinement step leads to unpredictable behavior of the TZS algorithm. The TZS only ends when the Refinement does not find a better result through the current iteration.

The Raster step is a sub-sampled full search that can be applied between the First Search and the Refinement. It is only used when the vector from the resulting block of the First Search is larger than the *iRaster* constant (*iRaster* is defined as 5 by default in the HM).

After the BMA finds the most similar block using the ME step, HEVC employs a refinement process over this block. This process is known as the FME (Fractional Motion Estimation), whereas it is common for authors to refer the ME at integer positions as the IME (Integer Motion Estimation). The HEVC standard employs the interpolation of samples in order to generate additional blocks at fractional positions for new comparisons. The HEVC uses FIR (Finite Impulse Response) filters with seven or eight taps to calculate both samples at quarter-pixel or half-pixel positions, which depends on the calculated position.

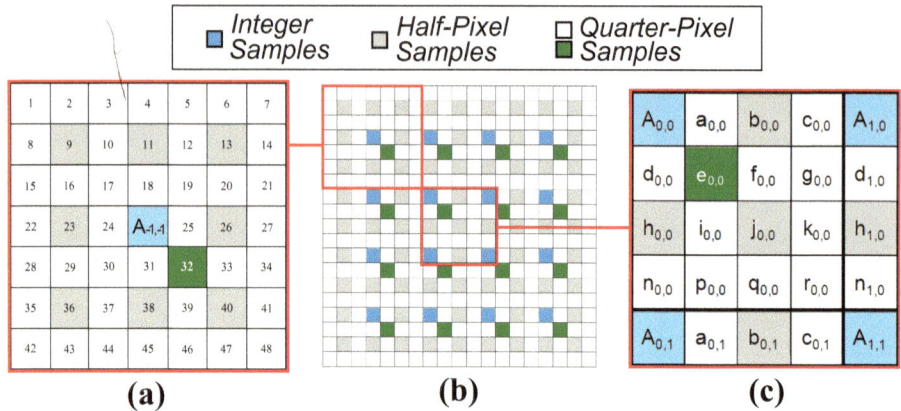

Fig. 2.18 Fractional motion estimation: **a** First samples of the 48 fractional blocks; **b** A 4×4 block; **c** Samples at integer and fractional positions (Adapted from Afonso et al. (2016))

The filter inputs can receive samples at integer positions or fractional positions previously calculated, depending on the position. Furthermore, the data dependence also varies according to the position to be calculated, i.e., the calculation of a specific sample at a fractional position depends on samples at half-pixel or quarter-pixel positions.

After the interpolation, a search-and-comparison process using samples at half-pixel and quarter-pixel is performed (Afonso et al. 2016). The search using the fractional samples occurs around the block with the best result considering integer-pixel positions. By default, in the HEVC reference-software implementation, a search with the eight blocks composed of half-pixel positions is performed firstly in the FME, and after that, a search with the eight blocks around the best match of half-pixel blocks is performed using quarter-pixel positions.

Figure 2.18 represents the samples at integer positions (blue squares and uppercase letters), as well as the samples at fractional positions (non-blue squares) for the interpolation process based on the HEVC standard. In Fig. 2.18b, a 4×4 block is represented (due to the space limitation). When fractional samples are generated, 48 new fractional blocks are formed for a new comparison, as presented in Fig. 2.18. In Fig. 2.18a, the values from 1 to 48 inside the squares represent the first sample of each new fractional block. The gray squares represent the half-pixel samples, and the white squares represent the quarter-pixel samples. In Fig. 2.18c, the fractional samples are detailed in lowercase letters. As an example, a fractional block with quarter-pixel precision is highlighted in green in Fig. 2.18b. It is important to note that the number of new blocks for comparison (48 fractional blocks) does not depend on the PU size.

Fifteen equations are used to calculate the fractional positions (ITU-T 2015) based on the FIR filters with 7-taps or 8-taps. The fractional positions $a_{0,0}$, $b_{0,0}$, $c_{0,0}$, $d_{0,0}$, $h_{0,0}$, and $n_{0,0}$ are calculated from the luminance values at integer positions. The calculation

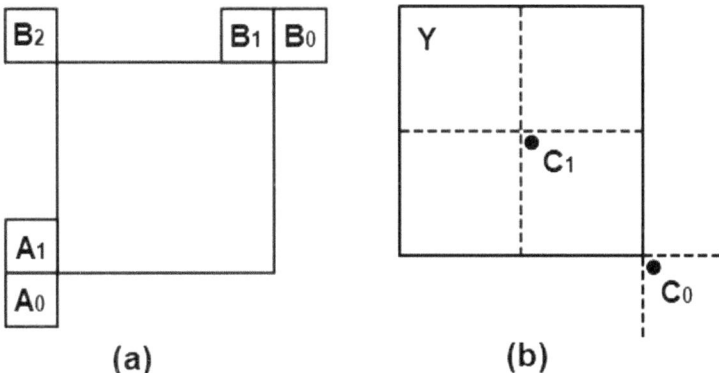

Fig. 2.19 Merge candidates: **a** spatial-neighboring candidates; **b** temporal-neighboring candidates

for determining the fractional positions $e_{0,0}$, $f_{0,0}$, $g_{0,0}$, $i_{0,0}$, $j_{0,0}$, $k_{0,0}$, $p_{0,0}$, $q_{0,0}$, and $r_{0,0}$ requires values of the positions $a_{0,i}$, $b_{0,i}$ and $c_{0,i}$ previously calculated, where i varies from -3 to 4 in the vertical direction (ITU-T 2015). It is important to notice that, during the interpolation process, some samples around the block are used to calculate the fractional samples. Since the filter inputs require seven or eight samples, a border of samples is needed to calculate the fractional samples located at the borders of the blocks.

Merge and Skip Modes

The HEVC, as well as the H.264/AVC, uses the Skip mode in the case of the PU being encoded is very similar to the collocated block in the reference frame. The main characteristic of the blocks encoded with Skip mode is that no information is transmitted by the encoder to represent the PU. In the HEVC, the concept of Skip mode is incremented through the Merge mode. In the Merge mode, the motion parameters of the PU being encoded are obtained from the neighboring spatial and temporal PUs previously processed, as presented in Fig. 2.19.

The main difference between the Merge mode and the Skip mode is that the motion parameters must be transmitted for each PU, including motion vectors and a reference frame index. When a PU is predicted with Merge mode, an index pointing to a Merge candidate list is analyzed from the bitstream and so, used to recover the motion information. The Merge candidate list construction is specified in the HEVC standard document (ITU-T 2015) and this list has a maximum number of candidates considering the blocks at neighboring positions of the block being encoded. Considering the spatial-neighboring blocks, until four candidates are selected among five possible candidates located in specific positions. In the case of the temporal-neighboring blocks, only one candidate can be selected among the two possible candidates. Since the decoder assumes a constant number of Merge candidates for each PU, additional candidates are generated when the number of candidates does not reach the maximum number signalized in the Slice Header.

The Skip mode can be applied to the 2Nx2N PU partition, i.e., to four block sizes, while the Merge mode can be applied to any one of the 24 possible PU sizes. More details about the coding tools employed in the HEVC standard can be obtained in (Sullivan et al. 2012; McCann et al. 2014), and in HEVC standard document (ITU-T 2015).

2.1.3   3D-HEVC Dependent-View Encoding

This section presents the novel encoding tools introduced by the 3D-HEVC encoder to better deal with dependent-view encoding.

2.1.3.1   Disparity-Compensated Prediction (DCP)

The Disparity Estimation (DE) technique is used in the 3D-HEVC Inter-view prediction to exploit inter-view redundancies. This technique was introduced by the MVC, and it uses a concept quite similar to the ME, widely used in the current video coding standards, such as the H.264/AVC and HEVC. The difference is that the DE uses previously processed neighboring views as reference frames, and the position is called disparity vector in this case.

Although DE and ME have many similarities, these techniques also present some important differences. Whereas ME explores the temporal redundancy of the same view, i.e., similarities between the temporally neighboring frames, DE explores the redundancy between different views, both in the texture and depth maps, considering the same time instant (the same Access Unit).

Figure 2.20 presents examples of the predictions based on DE and ME techniques. In the 3D-HEVC documents, these predictions are called DCP (Disparity-compensated Prediction) and MCP (Motion-compensated Prediction), respectively.

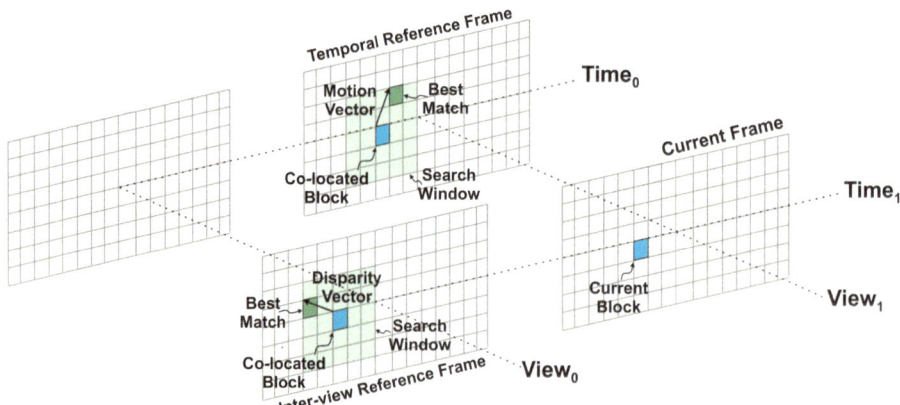

Fig. 2.20 Examples of MCP and DCP predictions

Another important difference between ME and DE is in the process used to generate the vectors. In general, ME generates vectors with reduced values for the block matching, which allows smaller search areas in the reference frames and, consequently, lower complexity in the process. Furthermore, in the ME, the search performed around the position of the collocated block in the reference frame tends to present a good result. On the other hand, DE tends to generate vectors with higher values than the ME for block matching. This way, searches around the position of the collocated block can result in an increased search window to reach satisfactory coding efficiency. The bigger the search window, the higher the computational effort needed for the encoding process. The 3D-HEVC coding tools were designed to be more efficient when the views are aligned in 1D linear and coplanar arrangements (Tech et al. 2016) and the camera arrangement used to record the videos follows a horizontal displacement (Müller et al. 2014; Tech et al. 2016). However, by default, the 3D-HEVC Reference Software (3D-HTM—3D-HEVC Test Model) defines a vertical search range equal to 56 and the TZS algorithm as the BMA for the DE, i.e., it performs a 2D search in a scenario where only 1D displacement is expected, demanding unnecessary computation by using an inappropriate algorithm for DE prediction. This book proposes some heuristics to exploit this issue in Chap. 5.

Given the similarities between DE and ME previously mentioned, these steps share the Motion Compensation (MC) step. It is important to notice that the DE is applied only in the dependent views.

In 3D-HEVC, as two different Inter predictions can be performed, the Inter-frames and the Inter-view predictions, two types of reference frames are included in two different reference lists (called L0 and L1). For the MCP, Inter-temporal reference frames are included and, for the DCP, Inter-view reference frames are included (previously processed frames that bellows to the same Access Unit). More details about the steps used to construct the reference lists can be found in (ITU-T 2015; Chen et al. 2015).

Despite the DE increasing the coding efficiency, DE competes with the ME in the encoding process. Since the ME presents better prediction than the DE in most cases, the DE is used in smaller regions of the frame. This way, the DE is mainly used to predict occluded regions due to the temporal motion (Müller et al. 2013).

Also, several techniques applied in the 3D-HEVC can use the obtained disparity vector to improve coding efficiency. More details about these techniques used in the 3D-HEVC can be found in Chen et al. (2015).

2.1.3.2   Modification in the Construction of the Merge Candidate List

The process used to construct the Merge Candidate List (MCL) is modified in the 3D-HEVC when compared to the HEVC. This modification aims to obtain improved efficiency in the prediction with a motion vector in the dependent views since the correlation of the information between the views can also be exploited. The maximum number of candidates in the MCL is increased to six in 3D-HEVC. Additional candidates are included in the MCL list considering the high probability of two associated blocks in the BV and the dependent views have the same motion information since these two views represent the same scene captured with synchronized cameras, only under different perspectives. These candidates consider the motion information of previously encoded blocks in the reference view and the disparity information that points to the Inter-view block in the reference view. The motion information of one of the dependent views can be predicted through the BV previously encoded since a disparity vector linking these two associated blocks is known. The steps to construct the MCL list for the dependent views of texture in the 3D-HEVC are described in detail in Chen et al. (2015).

2.1.3.3 Depth-Based Block Partitioning (DBBP)

Depth-Based Block Partitioning (DBBP) consists of a mode that derives an arbitrary partition for a given collocated texture block based on a Binary Segmentation Mask (BSM) calculated from the correspondent depth map. The two partitions obtained from this process are similar to the foreground and the background images, in which a merge step based on the BSM is applied. The process of generation of the masks starts with getting the disparity vector in order to identify the correspondent depth-map block in the reference view that is the same size of the current texture block. After the depth-map block is found, a threshold value is calculated based on the average value of the samples inside the correspondent depth-map block. This threshold value is used together with the depth values to generate the BSM $m_D(x,y)$. This process is relatively simple. If the depth value located in the (x,y) coordinates is higher than the threshold value, the $m_D(x,y)$ is equal to 1 and, otherwise, $m_D(x,y)$ is equal to 0. Figure 2.21 depicts the process used for the generation of the masks.

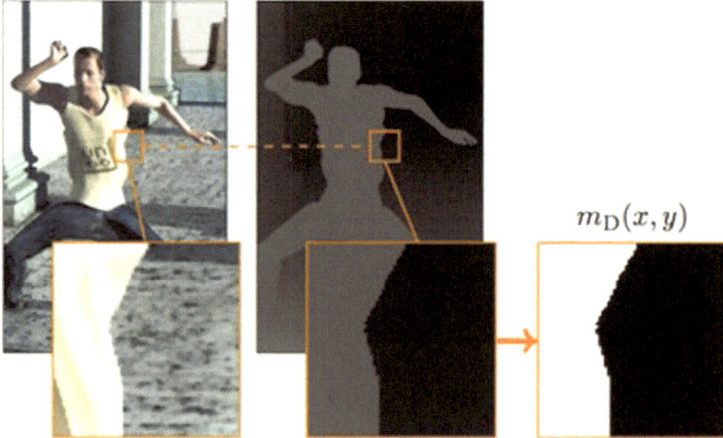

Fig. 2.21 Detailing of the mask generation using DBBP and the sequence *Undo_Dancer* (Müller et al. 2014; Chen et al. 2015, p. 29)

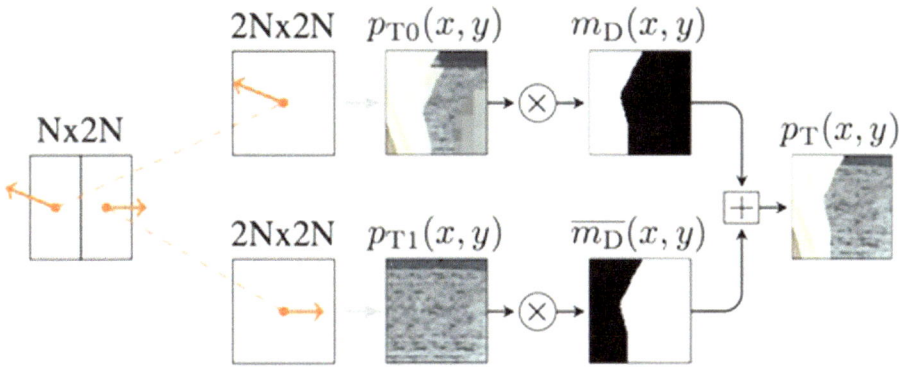

Fig. 2.22 Merging process used in the DBBP (Chen et al. 2015, p. 29)

In the case of BSM, two motion parameters are estimated in the encoder using a conventional ME and SAD as the similarity criterion. As one can observe in Fig. 2.22, a 2N × 2N block predicted using MCP is performed for each one of the two decoded motion parameters. In the following, the resultant prediction signals p_{T0} (x,y) e p_{T1} (x,y) are combined using the mask m_D (x,y). The two sets of motion parameters are predicted considering that the current block is encoded with two PUs of 2N × N or N × 2N sizes. Finally, the output p_T (x,y) is equal to p_{T0} (x,y), if m_D (x,y) is equal to 1. Otherwise, the output p_T (x,y) is equal to p_{T1} (x,y). It is important to notice that the DBBP mode is only applied to the 2N × N and N × 2N PU partitions, and to the 64 × 64, 32 × 32, and 16 × 16 CU-depth levels.

2.1.4   3D-HEVC Depth-Map Encoding

The depth-map encoding uses the same base concepts used in the coding of texture videos, such as the Intra-frame prediction, MCP prediction, DCP prediction, and Transform-based coding. However, some tools are modified, other tools are disabled, and novel encoding tools are introduced.

Due to the fact that one of the main aspects related to the depth maps is the presence of sharp edges, the interpolation filters of 7 and 8 taps used in the HEVC MC (samples with quarter-pixel precision) can produce artifacts when applied to the sharp edges. These artifacts can generate significant errors in the synthesized views. In order to avoid these errors in the synthesis process and reduce the encoder/decoder complexities, the MCP and DCP predictions are modified in the Inter-frames and Inter-view predictions, where the fractional prediction is not allowed.

Also, the in-loop filters used in HEVC have been developed to target the encoding of texture videos, which makes these filters inefficient in the depth-map encoding. This way, all in-looping filters (Deblocking Filter, and Sample-adaptive Loop Filter) are disabled in the depth-map encoding. Hence, the complexities of the encoder and the decoder can also be reduced.

The size of the Merge Candidate List for depth maps is also extended in one position, as in the texture encoding.

Since the HEVC Intra-frame prediction encoding tools are not so efficient in dealing with the sharp edges, nor with very homogeneous regions, 3D-HEVC introduces novel encoding tools along with the ones inherited from HEVC to enhance the efficiency of depth maps in the Intra-frame prediction. These novel encoding tools are explained in detail in the following sub-sections.

2.1.4.1   Depth Modeling Modes (DMM)

As previously mentioned, depth maps have some specific characteristics, such as the presence of sharp edges (representing the edges of the objects) and large areas with practically constant values or values with smooth variation (representing the inner parts of the objects).

The Intra-frame prediction and the transform-based encoding of the HEVC are efficient in encoding image regions with smooth variation. Therefore, the 35 Intra modes used in HEVC (33 directional modes, DC mode, and Planar mode) were maintained in 3D-HEVC to efficiently encode those regions. However, the use of those modes can generate artifacts by encoding sharp edges. Such artifacts result in significant errors in the process used for generating the synthetic intermediate views. To deal with this problem, new Intra modes called Depth Modeling Modes (DMM) were proposed for coding depth maps in 3D-HEVC.

Two DMMs are integrated with the conventional Intra-frame prediction used in HEVC. The DMMs generate a residue that represents the difference between the approximation of the depth map used in the encoding process and the original depth map. This residue can be transmitted through a transform-based coding, as in the Intra modes inherited from the HEVC. In these two new DMMs, each block of a given depth map is approximated using a model that partitions the internal area of the block into two regions represented by two constant values. Therefore, two information are needed: (i) the information that allows the identification of what partition the samples belong (called partition information); (ii) the constant value of the samples that belong to each region (called region value information). The information that corresponds to the value applied to each region is indicated by a CPV (Constant Partition Value). For a given partition, the better approximation is obtained by using the average value of the original depth values for the correspondent region as the CPV.

The DMMs differ in the segmentation of the block into two regions. These segmentations can be done using *Wedgelets* and *Contours*.

In the 3D-HEVC development, four Depth Modeling Modes were proposed for the Intra-frame prediction, but only two of those modes remain in the current version of the

Fig. 2.23 Wedgelet partition
of an 8 × 8 depth-map block

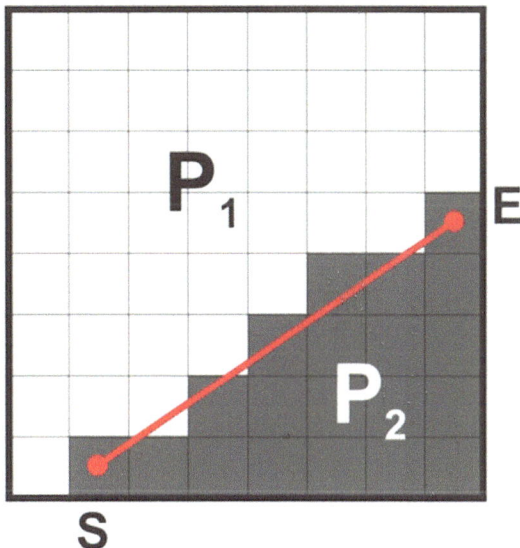

3D-HTM. These two modes are called: (i) DMM-1 or EWS (Explicit Wedgelet Signaling); and (ii) DMM-4 or ICPCP (Inter-component-predicted Contour Partitioning). Both DMM-1 and DMM-4 modes support 4 × 4, 8 × 8, 16 × 16 and 32 × 32 block sizes. Additional information about both DMM-1 and DMM-4 is given in the following.

DMM-1: EWS (Explicit Wedgelet Signaling)

When the DMM-1 is applied, the block is divided into two regions using a straight line, as presented in Fig. 2.23, which considers the partition of an 8 × 8 block as an example. The two regions are called P_1 and P_2. The start point of the straight line is represented by S, whereas the end point of this line is represented by E. These two points are always located on two different external borders of the block.

The DMM-1 mode finds the best match for each evaluated block using wedgelet partitions. The encoder performs a search over a set of wedgelet partitions using the original depth values of the current block as the reference. The wedgelet partition that presents a minimal distortion with the original depth signal is selected, and the correspondent partition information is transmitted in the bitstream. Each wedgelet partition is decided using the start and the end positions of the segmentation line along the borders of the current block.

The 3D-HTM uses different lists to store the wedgelets evaluated by the DMM-1 mode according to the block size. These sets of wedgelets are generated before the coding process, and all the wedgelets inside a specific list are evaluated during the prediction of their respective block size. The generation of the wedgelets occurs in two steps, one initial step followed by a refinement step. In the first step, all wedgelets between two adjacent and between two opposite borders are created and stored for each list according to the block size. This process starts with two adjacent borders and, as soon as these wedgelets are obtained, this set of wedgelets is rotated in order to generate all wedgelets considering adjacent borders. The generation of wedgelets for opposite borders is similar (Zhang et al. 2014; Ikai et al. 2015). It is important to notice that the generation of wedgelets for different block sizes applies different sampling to determine the start and end points. The 16×16 and 32×32 block sizes consider start and end points every two samples along the borders, whereas the 4×4 and 8×8 block sizes consider start and end points every sample along the borders, i.e., 4×4 and 8×8 block sizes consider all possible wedgelets. However, after the generation of the wedgelets among adjacent and opposite borders, the 3D-HTM implementation removes the wedgelets with the same start and end points or the ones with higher similarity.

Figure 2.24 presents the algorithm applied to the DMM-1 mode according to the 3D-HTM implementation. For each DMM-1 partition to be tested, the associated wedgelet pattern is searched in the Wedgelet Memory. Basically, each storage wedgelet consists of a binary mask and, an average value to be used in each region (CPV value) is computed based on this binary mask from the original depth block. After the calculation of the average values, the predicted block is derived where each region assumes the respective CPV value previously calculated. The following step consists of the distortion computation in which the lowest distortion is updated until all initial patterns defined in 3D-HTM are evaluated. When all initial patterns are evaluated, the wedgelet that presents the lowest distortion in previous steps is used for a Refinement Step, where additional patterns are tested. The refinement step consists of generating eight new wedgelets patterns to be evaluated around the previously selected wedgelet. These new wedgelets are generated apart from one sample of the previously selected wedgelet covering the eight possible cases. After all evaluations, the selected wedgelet and the residues are delivered by the Residue Computation.

The optimal results for the DMM-1 are reached by testing all possible wedgelets combining start and end points from the different block borders, which requires a high computational effort. As previously mentioned, the 3D-HTM implementation applies some restrictions to the start and end positions of the wedgelet partitions in order to reduce the number of wedgelet partitions to be evaluated, the storing of wedgelets in the Wedgelet Memory, and consequently, the complexity associated with the DMM-1 mode. The reduced number of wedgelet partitions used for comparisons, according to the block size, is presented in Table 2.1. Still, DMM-1 is a bottleneck in both processing and memory since a large number of wedgelet patterns are processed and 183,264 bits are required to store all wedgelets patterns (Zhang et al. 2014; Ikai et al. 2015) of all block sizes supported by the encoding tool, according to the 3D-HTM implementation.

On the decoding side, the predicted block is reconstructed using the partition information transmitted in the bitstream. In DMM-1, the partition information is not predicted, therefore this mode is called Explicit Wedgelet Signaling. Figure 2.25 shows the DMM-1

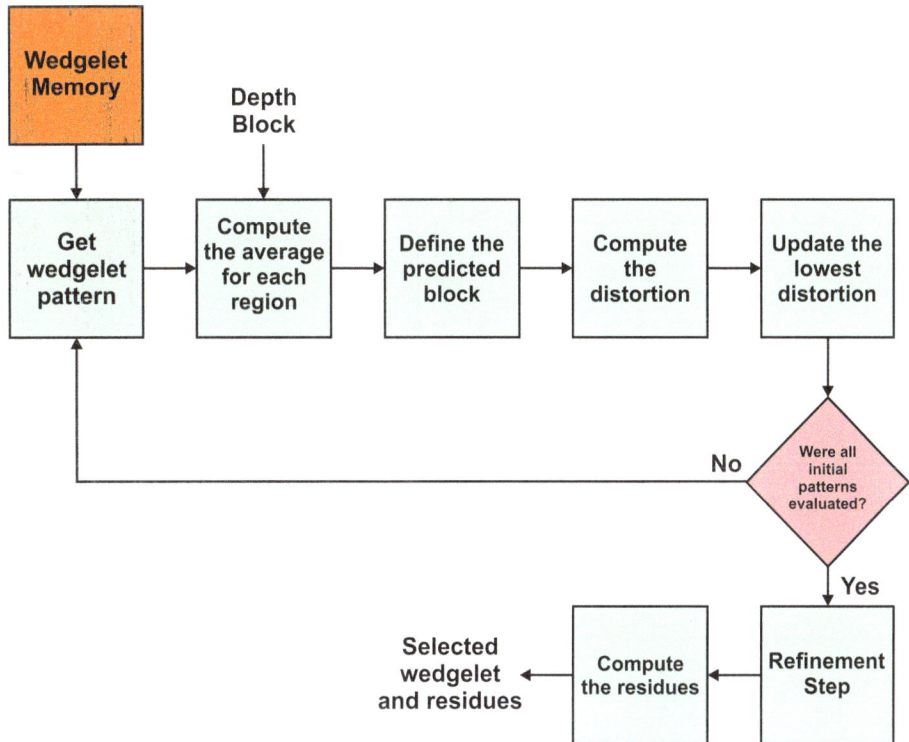

Fig. 2.24 DMM-1 encoding algorithm according to the 3D-HTM implementation

Table 2.1 Number of DMM-1 wedgelets according to the 3D-HTM implementation (Ikai et al. 2015)

Block size	Total of possible wedgelets
4×4	86
8×8	802
16×16	510
32×32	510

decoding process for a 4×4 depth-map block, as an example. For a given depth map block, the decoding algorithm requires the selected number of the wedgelet pattern, the CPV for each encoded region, and the residual values. While the DMM-1 encoding process evaluates an expressive amount of wedgelet patterns, which results in many memory accesses, the decoding process only requires access to the wedgelet indexed by the pattern number transmitted to the decoder. Then, the wedgelet pattern that was selected in the encoding process is retrieved. After that, the CPV values are mapped into this DMM-1

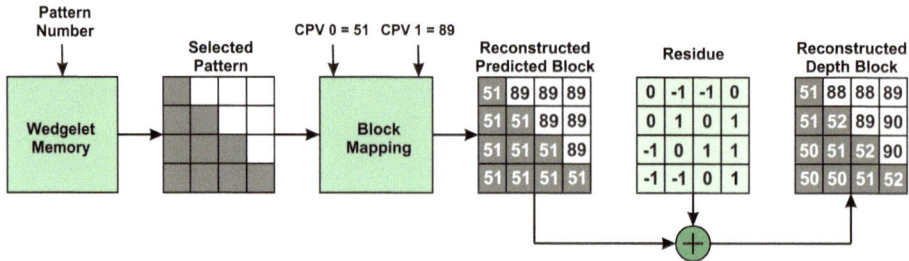

Fig. 2.25 Example of the DMM-1 decoding process for a 4 × 4 depth-map block

Fig. 2.26 Contour partition of an 8 × 8 depth-map block

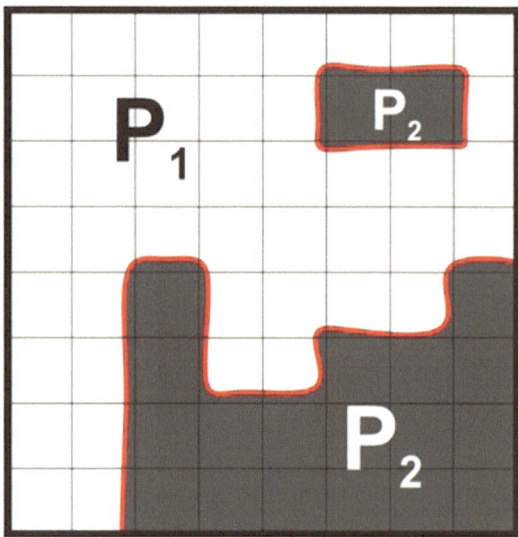

pattern to generate the reconstructed predicted block. Finally, the residue values are added to the reconstructed predicted block to generate the reconstructed depth map block.

DMM-4: ICPCP (Inter-Component-Predicted Contour Partitioning)

In the case of the Contour partition, the segmentation of the two regions cannot be easily described by a geometric function, since the regions P_1 and P_2 can have arbitrary forms and several parts. An example of a Contour partition can be observed in Fig. 2.26. The partition scheme considering the contour partition is individually obtained for each block from the reference block, i.e., no list can be used.

The DMM-4 mode consists of the prediction of the Contour partition (see Fig. 2.26) from a texture reference block using Inter-component prediction. The reconstructed luminance signal of the collocated block in the associated texture frame is used as the reference. The DMM-4 is based on the Rate-Distortion (RD) cost computation, i.e.,

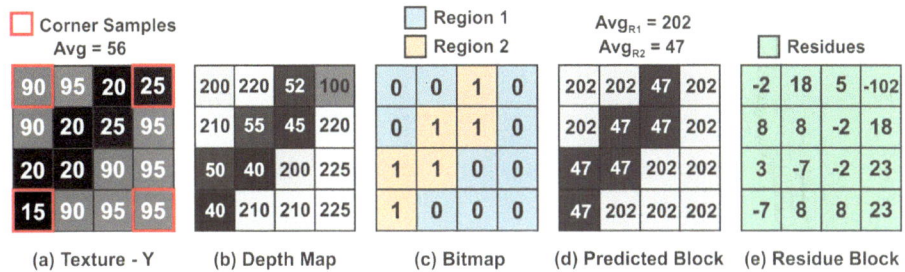

Fig. 2.27 Example of the DMM-4 coding process for a 4×4 depth-map block

according to the RD-cost results, the selection among DMM-4 and the other intra prediction modes is made based on the lowest RD cost.

As previously explained, by using Contour partition along the DMM-4, the depth block to be predicted is divided into two regions (or partitions) in which discontinuous regions may belong to each partition, which is not possible with the DMM-1. This division is done using a threshold value from the texture samples of the collocated block inside the texture picture associated with the depth map. From the threshold value, a contour of the objects is constructed as presented in Fig. 2.27. This figure shows an example of prediction with the DMM-4 mode considering a 4×4 block size. As one can observe in Fig. 2.27a, the threshold is obtained from the average using the four corner samples presented in the collocated texture reference block (considering the luminance channel— Y), resulting in an average value equal to 56 for this example. The threshold value is then used to divide the depth block represented in Fig. 2.27b into two regions from a bitmap. Those samples that present values higher than the threshold value are identified as the samples that belong to a region 1, whereas the samples with values smaller than the threshold are associated to the region 2 in the Contour partition.

Figure 2.27c shows the bitmap obtained from the depth block given as an example in Fig. 2.27b. Note that the depth-block values are compared with the threshold value obtained from the texture in order to define if the bitmap receives *0* or *1*. Once this is done, the DMM-4 mode calculates the arithmetic average of all depth block samples of each region from the bitmap to determine the value that will entirely represent each region. This way, the bitmap can be replaced with the average of the partitions to generate the predicted block, as presented in the example shown in Fig. 2.27d, where the two regions result in average values equal to 202 and 47. Finally, the DMM-4 calculates the difference between the original and predicted samples and delivers the residues, as shown in Fig. 2.27e.

2.1.4.2   DIS (Depth Intra Skip) Mode

As previously mentioned, one of the main aspects of depth maps is the presence of many regions with similar values and, frequently, the same value. In most cases, small variations

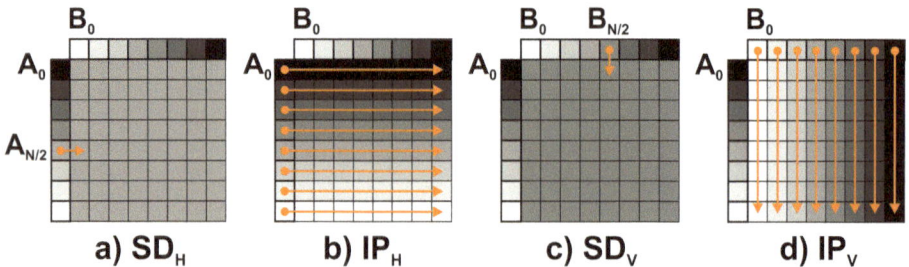

Fig. 2.28 Depth Intra Skip sub-modes: **a** SD$_H$; **b** IP$_H$; **c** SD$_V$; **d** IP$_V$ (Adapted from Afonso et al. (2017))

of the sample values of homogeneous regions insignificantly affect the quality of the synthesized views. With this in mind, the DIS mode was introduced in the 3D-HEVC. This coding tool avoids the residual coding of homogeneous areas in depth maps, drastically reducing the amount of data required to represent those regions.

DIS presents four prediction sub-modes as presented in Fig. 2.28: (i) Horizontal Single Depth (SD$_H$); (ii) Horizontal Intra Prediction (IP$_H$); (iii) Vertical Single Depth (SD$_V$); and (iv) Vertical Intra Prediction (IP$_V$). SD modes consider only one depth sample of previously processed neighboring blocks to perform the prediction of the current block, i.e., the whole block is predicted with the same value. While the SD$_H$ considers the sample from the left neighboring column located in the position A$_{N/2}$, SD$_V$ uses the sample from the above neighboring row located in the position B$_{N/2}$, where N represents the block size. Note that Fig. 2.28 uses an example with 8×8 blocks. When a spatial neighboring sample is not available for the SD modes, the middle value of the depth range is used for the whole block (e.g., 128 for 8 bits). The IP$_H$ and IP$_V$ prediction modes apply the horizontal and vertical modes commonly used in the HEVC. It consists of copying the reference samples in horizontal and vertical directions, respectively.

By default, DIS uses SVDC (Synthesized View Distortion Change) as the distortion metric to decide the best mode to encode a given block based on RDO (3D-HEVC Reference Software 2024). SVDC uses rendering functionalities, as will be explained in Sect. 2.1.5.1.1.

DIS is applied over blocks at the level of CUs, and it can encode any of the possible CU sizes (8×8, 16×16, 32×32, and 64×64 blocks). An index is used to identify the prediction mode of the candidate list of the CU encoded using DIS.

2.1.5   Control of the 3D-HEVC Encoder

The Rate Distortion Optimization (RDO) technique is used in the decision mode to measure the cost of the different encoding modes and parameters applied to the candidate blocks in the encoding process. The mode or parameter that returns the smallest cost according to Eq. (2.18) is selected.

$$J = D + \lambda.R \qquad (2.18)$$

In Eq. (2.18), J represents the cost, D represents the distortion obtained by encoding a given block with a specific mode, R represents the number of bits needed to represent the block in this specific mode, and λ is the Lagrangian multiplier obtained based on the QP. Methods such as SSD (Sum of Squared Differences), SAD, or SATD are applied to obtain the distortion between the original and the reconstructed sample values.

For the depth-map encoding, the same decision process is applied. However, the distortion is measured using a modified process, and it also considers the distortion of the synthesized views, which increases the coding efficiency. This modification is necessary since the geometric information of the depth maps is indirectly explored during the rendering process and the decoded depth map is not visible during this process.

2.1.5.1 View Synthesis Optimization (VSO)

Occlusions and disocclusions avoid a bijective mapping of depth-map areas with distortion regions in the synthesized view. This occurs when the block (or part of this block) is occluded in the synthesized view. Hence, an exact mapping between the depth-block distortion and the associated distortion in the synthesized view is not possible by using only the information provided by the current block. To measure the distortion of the synthesized views, the Synthesized View Distortion Change (SVDC) metric can be applied in RDO.

Synthesized View Distortion Change (SVDC)

The Synthesized View Distortion Change (SVDC; Chen et al. 2015) allows an exact measuring of the synthesized-view distortion that considers the occlusions and disocclusions. As presented in Fig. 2.29, the distortion of the synthesized view as a function of the changes inside the depth block B is calculated considering the depth information outside block B. For that, SVDC is defined as the difference in distortion ΔD of two texture synthesized views s'_T and $\tilde{s}'_T$ and it is mathematically represented by Eq. (2.19). s'_T represents a texture rendered from the depth map s_D which consists of depth information reconstructed by blocks previously encoded and original depth information for the remaining blocks. $\tilde{s}'_T$ represents a texture rendered from the depth map $\tilde{s}_D$. In this case, the reconstructed depth values from the current encoding mode are used considering the current block B. The distortions $\mathbf{D}$ and $\tilde{\mathbf{D}}$ are calculated using the SSD metric. For calculating the SSD, $s'_{T,Ref}$ represents a reference texture in the initialization rendered using the original video and the depth information. Finally, the difference between $\mathbf{D}$ and $\tilde{\mathbf{D}}$ is represented by the $\Delta\mathbf{D}$, and it measures the depth distortion. The VS blocks in Fig. 2.29 represent the Views Synthesis steps. I represents the set of all samples in the synthesized view.

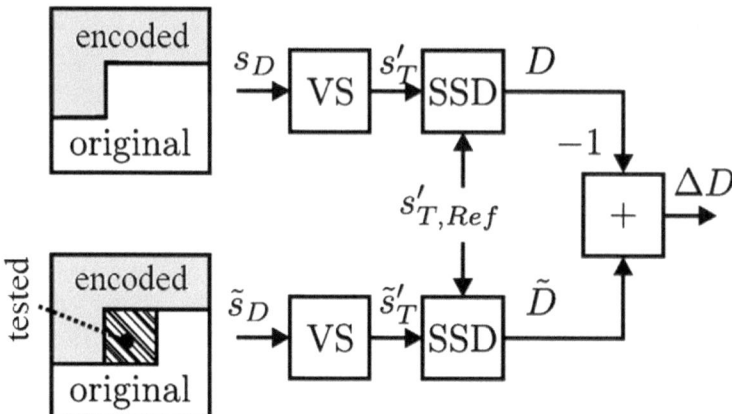

Fig. 2.29 SVDC definition related to the distorted depth information of the block B (represented by the hatched area) (Adapted from Chen et al. (2015, p. 44))

$$SVDC = \Delta D = \tilde{D} - D = \sum_{(x,y)\in I} \left[\tilde{s}_T(x, y) - s'_{T,Ref}(x, y)\right]^2 - \sum_{(x,y)\in I} \left[s_T{}'(x, y) - s'_{T,Ref}(x, y)\right]^2$$

$$(2.19)$$

The SVDC calculation requires a rendering process during the encoding. Given that the required computational effort is a critical factor in the encoding, an alternative method is used. This approach uses fast rendering in order to avoid re-rendering of synthesized views for each distortion calculation. That method supports the base functionalities used in the processing of view synthesis algorithms of the most rendering approaches, such as the Sub-Sample Accurate Warping, Hole Filling, and View Blending. More details about the SVDC metric can be obtained in Chen et al. (2015).

2.2   3D-HEVC Related Works

Since the 3D-HEVC was released in 2015, papers have been published targeting efficient hardware designs for the new proposed encoding tools or specifically focusing on the 3D-HEVC challenges. However, since the HEVC tools are also required in the 3D-HEVC implementations, previously published works targeting the HEVC can also be used in 3D-HEVC implementations with some adaptations, as will be discussed in the following.

2.2.1 HEVC Related Works

All prediction modes available at the HEVC Intra-frame prediction are also used in the 3D-HEVC. Then, the hardware designed targeting the HEVC Intra-frame prediction can be used in 3D-HEVC implementations. Some works like (Zhang et al. 2019; Huang et al. 2016; Min et al. 2017; Correa et al. 2017; Palomino et al. 2012; Pastuszak et al. 2016a) should be considered when designing dedicated hardware for a complete 3D-HEVC codec. These architectures can be adapted to include the new prediction modes defined by the 3D-HEVC.

The hardware designed targeting the HEVC Inter-frames prediction (motion estimation and motion compensation) can be used in the 3D-HEVC Inter-frames prediction. Since the Inter-frames and Inter-view predictions have very similar behavior, solutions designed targeting the HEVC Inter-frames prediction can be replicated to work in the 3D-HEVC Inter-view prediction with adaptations. Thus, works (Tseng et al. 2019; Xu et al. 2018; Lung et al. 2019; Alcocer et al. 2019; Shi et al. 2020; Jou et al. 2015; Gogoi et al. 2021; Perleberg et al. 2018; Sanchez et al. 2015b; Kalali et al. 2018; Fan et al. 2018) can be used to design a 3D-HEVC codec.

The hardware designed targeting other HEVC encoder modules: forward and inverse transforms, forward and inverse quantization, entropy coding, and in-loop filters, can also be reused in the 3D-HEVC context with minor adaptations, including the works (Kalali et al. 2014; Lee et al. 2016; Braatz et al. 2018; Goebel et al. 2016). The HEVC entropy coding, due to its natural control flow behavior, is more complex to parallelize, then designs of high throughput hardware are not easy to be found in the literature, but a few works (Ramos et al. 2018; Ramos et al. 2016; Kim et al. 2015b; Zhang et al. 2018) can be used in this scenario. Finally, the published hardware designs targeting the HEVC in-looping filter also can be reused in 3D-HEVC implementations, including works (Fang et al. 2015; Shen et al. 2016; Cho et al. 2015; Zhou et al. 2016). However, none of these solutions is capable of handling depth map coding and most of them are not able to reach the required throughput to process 3D videos in real-time.

In turn, (Lee et al. 2017) implements depth map coding using the baseline HEVC standard. It proposes an Adaptive Search Window (ASW) that explores the temporal correlation between depth map and texture. The correlation is used to predict the probable range of movements of an object throughout consecutive frames, which results in an encoding time reduction of up to 53% in comparison to TZS.

Moreover, such solutions targeting the HEVC do not feature optimizations considering the runtime 3D-HEVC behavior and processing/memory requirements. Multiple related works proposing memory and energy-aware hardware designs are available in the literature, but focus on previous 3D video-coding standards, as will be explained in the next section.

2.2.2   H.264/AVC and MVC Related Works

Several related works proposing memory and energy-aware hardware designs for ME and DE have been published lately. In the following, some of the most prominent works targeting Inter-frames and Inter-view predictions for previous video coding standards are discussed.

Aiming to reduce the number of block matching operations, many solutions propose Adaptive Search Window (ASW) algorithms. The work (Jia et al. 2013) proposes applying the exhaustive full search (FS) algorithm within a diamond-shaped ASW that is dynamically resized at the frame level. When implemented on an H.264/AVC encoder, the solution reached an encoding time reduction of 80%. The authors in (Kim et al. 2014) propose an ASW algorithm and use a fast ME algorithm to accelerate the HEVC ME based on Graphics Processing Units (GPU). A scalable fast search algorithm suitable for a massively parallel GPU architecture focusing on MVC ME/DE is proposed by (Jiang et al. 2016). A hardware-oriented fast Integer ME (IME) algorithm is proposed by (Doan et al. 2017). It targets a future parallel ME design capable of processing all HEVC block sizes while reducing the computational complexity by 54.54%. None of the available proposals is able to exploit the tools provided by 3D-HEVC. Moreover, the solutions are not focused on real-time processing and neither consider memory-related issues.

A handful of solutions to reduce and control video on-chip memories are already available. The works (Chen et al. 2006; Tsung et al. 2016) are among the most widely used on-chip memory data reuse strategies, in which the SW is fully stored on-chip to reduce the accesses to the external memory. However, several samples within the SW are not used due to the adaptive characteristic of the current ME/DE BMAs. The work (Zatt et al. 2011a) proposes on-chip video memory architecture for ME/DE in MVC, in which an application-aware power management scheme based on a multiple-sleep state model is employed. In (Sampaio et al. 2013), an energy-efficient memory hierarchy for ME/DE on MVC was presented. The authors employed a *Reference Frames-Centered Data Reuse* scheme to avoid multiple search window retransmissions, leading to a reduced number of external memory accesses and, consequently, memory energy reduction. The on-chip video memory energy was reduced by employing a statistical power gating scheme and candidate block reordering. Therefore, up to 71% of external memory energy, 88% of on-chip memory static energy, and 65% of on-chip memory dynamic energy were saved when compared to the Level-C approach. Unfortunately, none of the described solutions considers depth maps and are unable to exploit motion and disparity characteristics to further reduce energy consumption.

A few complete systems for MVC have also been proposed. In Ding et al. (2010) the authors present an MVC encoder with low energy consumption able to encode four high definition views in real-time. In Zatt et al. (2011b), a run-time adaptive energy-aware ME/ DE architecture for the MVC standard is presented, which uses memory access and data prefetching techniques for jointly reducing the on-/off-chip memory energy consumption. The work (Zatt et al. 2011b) also proposes a run-time dynamically expanding SW to reduce the off-chip memory accesses and a power-gating scheme to reduce on-chip memory power dissipation. As a result, (Zatt et al. 2011b) provides a dynamic energy reduction of 82–96% for the off-chip memory and a leakage energy reduction of 57–75% for the on-chip memory in comparison to Level-C and Level-C+. The authors in (Choi et al. 2013) propose two MVC frame scheduling schemes considering MVC with depth information. These scheduling schemes avoid that the different encoding times of each channel reduce the coding efficiency.

2.2.3 MV-HEVC and 3D-HEVC Related Works

There exist prominent works in the literature proposing dedicated architectures and algorithmic solutions for the novel Intra-frame encoding tools of the 3D-HEVC.

To efficiently encode depth maps, the 3D-HEVC adopts new encoding tools, as previously discussed. In Intra-frame prediction, 3D-HEVC introduces: (i) Depth Intra Skip (DIS); (ii) Depth Modeling Mode-1 (DMM-1); and (iii) Depth Modeling Mode-4 (DMM-4). These tools increase the coding efficiency but also increase the required computational effort.

The works (Kim et al. 2015a; Conceição et al. 2016) present algorithm solutions to the 3D-HEVC DIS. The work (Kim et al. 2015a) proposes a fast mode decision for the Single Depth (SD) modes based on a decision criterion to detect smooth regions in depth maps. It early decides the use of SD modes by using statistics of smooth signals for depth intra modes and analysis of the distortion metrics. As a result, this work presents a 25.6% encoding-time saving at the cost of a 0.18% increase in the BD-Rate. The work (Conceição et al. 2016) proposes an Early Skip/DIS mode decision for 3D-HEVC to reduce the complexity of the depth-map coding process. This work is based on an adaptive threshold model that considers the Skip/DIS occurrences as a function of its generated rate-distortion cost. This solution reduces in 33.7% the depth complexity on average with a mean BD-rate increase of up to 0.409%. The work (Ahmad et al. 2015) presented complexity, and hardware architecture analyses for the implementation of Synthesized View Distortion Estimation, which is the distortion metric adopted in RDO to evaluate the encoding of blocks using the DIS tool.

The main works published in the literature that propose hardware designs for the 3D-HEVC depth modeling modes, DMM-1 and/or the DMM-4, are (Amish et al. 2019; Sanchez et al. 2014b; Sanchez et al. 2016; Sanchez et al. 2017b; Sanchez et al. 2018a; Sanchez et al. 2019). The work presented in (Sanchez et al. 2014b) implements the DMM-4 prediction mode for 8×8, 16×16, and 32×32 block sizes. The architecture was synthesized for an Altera Stratix V FPGA, and it is capable of processing HD 1080p videos (1920×1080 pixels) with five views at 31.39 frames per second (fps). The work presented in (Sanchez et al. 2016) implements the DMM-1 and DMM-4 prediction modes, which can be scaled for all block sizes. DMM-1 memory issues are not covered by this work. The architecture was synthesized for ASIC, and it is capable of processing HD 1080p videos at 30 fps considering one view. The work presented in (Amish et al. 2019) implements the three modes introduced in the 3D-HEVC extension: DIS, DMM-1, and DMM-4 considering all possible block sizes. A strategy to reduce the DMM-1 complexity was implemented, but memory issues were not covered. The developed architecture was synthesized for a Virtex 6 FPGA, and it can process HD 1080p videos at 30 fps, considering six views. The work (Sanchez et al. 2019) presents hardware designs for the encoder and decoder and for both DMM-1 and DMM-4 encoding tools of the 3D-HEVC. The encoder was designed using a simplification in the DMM-1 algorithm. The encoder architecture was designed in a scalable structure that supports different block sizes and reaches different throughputs according to the application requirements. The decoder was designed to share resources between DMM-1 and DMM-4 execution and support all available block sizes. Both architectures were synthesized for Standard Cell ST 65 nm and 28 nm technologies targeting the processing of 1920×1080 videos at 30 fps.

The works (Amish et al. 2019; Sanchez et al. 2014b; Sanchez et al. 2016) are not fully compliant with the 3D-HEVC standard. The works (Sanchez et al. 2014b; Sanchez et al. 2016) present differences in the method used to calculate the texture block average in DMM-4 besides the work (Amish et al. 2019) uses an alternative method to calculate the DIS. A DMM-1 decoder implemented in hardware and supporting the processing of high-resolution videos in real-time is presented in (Sanchez et al. 2018a). The work (Sanchez et al. 2017b) presents an architecture for the Simplified Edge Detector (SED) algorithm employed in 3D-HEVC depth-maps Intra-frame prediction.

Only one work was found in the literature focused on hardware design targeting the MV-HEVC. The work (Liu et al. 2017) presents a hardware design for the MV-HEVC focusing on the real-time decoding of multi-view videos in mobile devices.

2.2.4 Related Works Summary

Table 2.2 summarizes some important information related to the most prominent related works found in the literature. Even though there are related works (Tsung et al. 2016; Zatt et al. 2011a; Zatt et al. 2011b; Sampaio et al. 2013; Ding et al. 2010) that propose memory and energy-aware hardware systems for ME/DE of previous video coding standards (MVC), none of them focuses on 3D-HEVC. This means that many of them do not support depth map coding (or apply texture-only video codecs to encode depth maps) and, as a result, these works do not exploit the specific depth-map characteristics. Additionally, no solutions that fully exploit the MVD correlation space (spatial, temporal, disparity and inter-channel) to jointly reduce computation and memory-related energy consumption were found in the literature.

Table 2.2 Related works summary

Related work	Video-coding standard	Encoding tools	Depth-map processing	Memory implementation
Zhang et al. (2019)	HEVC	Intra-frame Prediction	No	Yes
Huang (2016)	HEVC	Intra-frame Prediction	No	Yes
Min et al. (2017)	HEVC	Intra-frame Prediction	No	No
Correa et al. (2017)	HEVC	Intra-frame Prediction	No	No
Palomino et al. (2012)	HEVC	Intra-frame Prediction	No	No
Pastuszak et al. (2016a)	HEVC	Intra-frame Prediction	No	No
Tseng et al. (2019)	HEVC	Inter-frames Prediction	No	No
Xu et al. (2018)	HEVC	Inter-frames Prediction	No	Yes
Lung et al. (2019)	HEVC	Inter-frames Prediction	No	No
Alcocer et al. (2019)	HEVC	Inter-frames Prediction	No	Yes
Shi et al. (2020)	HEVC	Inter-frames Prediction	No	Yes
Jou et al. (2015)	HEVC	Inter-frames Prediction	No	Yes
Gogoi et al. (2021)	HEVC	Inter-frames Prediction	No	Yes
Perleberg et al. (2018)	HEVC	Inter-frames Prediction	No	No
Sanchez et al. (2015b)	HEVC	Inter-frames Prediction	No	No
Kalali et al. (2018)	HEVC	Inter-frames Prediction	No	No
Fan et al. (2018)	HEVC	Inter-frames Prediction	No	No
Kim et al. (2014)	HEVC	Inter-frames Prediction	No	No

(continued)

Table 2.2 (continued)

Related work	Video-coding standard	Encoding tools	Depth-map processing	Memory implementation
Doan et al. (2017)	HEVC	Inter-frames Prediction	No	No
Lee et al. (2017)	HEVC	Inter-frames Prediction	Yes	No
Jia et al. (2013)	H.264/AVC	Inter-frames Prediction	No	No
Chen et al. (2006)	H.264/AVC	Inter-frames Prediction	No	No
Tsung et al. (2016)	MVC	Inter-frames and Inter-view Predictions	No	No
Zatt et al. (2011a)	MVC	Inter-frames and Inter-view Predictions	No	Yes
Zatt et al. (2011b)	MVC	Inter-frames and Inter-view Predictions	No	Yes
Sampaio et al. (2013)	MVC	Inter-frames and Inter-view Predictions	No	Yes
Ding et al. (2010)	MVC	Encoder	No	Yes
Kim et al. (2015a)	3D-HEVC	*Intra-frame DIS tool	Yes	No
Conceição et al. (2016)	3D-HEVC	*Intra-frame DIS tool	Yes	No
Amish et al. (2019)	3D-HEVC	**Intra-frame DIS, DMM-1, and DMM-4 tools	Yes	No
Sanchez et al. (2014b)	3D-HEVC	***Intra-frame DMM-4 tool	Yes	No
Sanchez et al. (2016)	3D-HEVC	***Intra-frame DMM-1, and DMM-4 tools	Yes	No
Sanchez et al. (2018a)	3D-HEVC	Intra-frame DMM-1 tool (decoder)	Yes	No

(continued)

Table 2.2 (continued)

Related work	Video-coding standard	Encoding tools	Depth-map processing	Memory implementation
Sanchez et al. (2019)	3D-HEVC	Intra-frame DMM-1, and DMM-4 tools (encoder and decoder)	Yes	Yes
Liu et al. (2017)	MV-HEVC	Decoder	No	No

[*]Algorithmic solution
[**]Not fully compliant with the 3D-HEVC standard (DIS encoding tool)
[***]Not fully compliant with the 3D-HEVC standard (DMM-4 encoding tool)

Considering the hardware designs (Amish et al. 2019; Sanchez et al. 2014b; Sanchez et al. 2016; Sanchez et al. 2018a; Sanchez et al. 2019) which are capable of dealing with depth-map processing using the 3D-HEVC (MVD approach), all works implemented Intra-frame solutions for specific encoding tools without consider general aspects related to the Intra-frame prediction nor some important issues related to the memory usage and energy consumption required by the encoding tools.

Given that the 3D-HEVC extension exploits the MVD concept, and the MVD concept can be considered state of the art in terms of an approach to deal with 3D-video coding, bringing innovations along with its depth-map processing and novel encoding tools, an investigation of the 3D-HEVC encoding tools and the development of algorithms and hardware architectures capable of processing multi-view 3D-videos in real-time are motivating and challenging activities.

This chapter presents the evaluations performed through the 3D-HEVC Reference Software, the 3D-HTM, and it is divided into three main sections. The first one presents 3D-HTM Time Profiling and Memory Analyses. The second section presents a statistical analysis based on the encoding tools considering both the CCO (Conventional Coding Order) and the FCO (Flexible Coding Order) configurations. These statistical analyses were performed with the goal of identifying the importance of each encoding tool and, this way, to propitiate the proposal of efficient complexity-reduction strategies for the 3D-HEVC encoder. The third one discusses the obtained results. It is important to note that the major part of the experiments performed for this book followed the test recommendations provided by the JCT-3V and used version 16.0 of the 3D-HEVC reference software. This experimental setup is shown in detail in Appendix A. There are a few exceptions from evaluations where the experimental setup or the 3D-HEVC reference software version is different from those mentioned above. In these specific cases, the differences are highlighted throughout the text and discussed.

57
V. Afonso et al., *Hardware Design for 3D Video Coding*, Synthesis Lectures on Engineering, Science, and Technology, https://doi.org/10.1007/978-3-031-80232-4_3

3.1   3D-HTM Time Profiling and Memory Analyses

The new coding tools introduced by 3D-HEVC to efficiently encode MVD content result in a higher complexity when compared to the HEVC. It is important to mention that the HEVC already has a high complexity when compared with previous standards. Therefore, the development of real-time 3D systems, which is the focus of this book, is a very challenging task, demanding dedicated energy-aware hardware solutions to enable real-time encoding, especially when portable devices are considered. The relevance of this research topic is placed in evidence due to the high 3D-HEVC complexity and its features.

The memory and computational effort required for both the novel depth 3D-HEVC intra prediction tools, such as the DMM-1 mode, and the 3D-HEVC ME/DE steps evidences the need for the development of complexity-reduction strategies and VLSI designs focusing on both 3D-HEVC Intra-frame prediction, and Inter-frames and Inter-view predictions.

The main challenges regarding memory, processing, and complexity for the 3D-HEVC predictions (Intra-frame, Inter-frames, and Inter-view) are discussed in the next two subsections.

3.1.1   3D-HEVC Intra-Frame Prediction

Figure 3.1 presents an analysis of 3D-HEVC Intra-frame prediction regarding the time profiling of 3D-HEVC intra coding. These results were obtained through evaluations using the 3D-HTM (3D-HEVC Test Model) reference software version 16.0 (3D-HEVC Reference Software 2024) under AI (All-Intra) encoder configuration (see Sect. 2.1.1.2), i.e., only Intra-frame prediction tools are available to encode texture and depth maps. The Common Test Conditions (CTC) for 3D-HEVC presented in Appendix A were adopted as the experimental setup.

Analyzing the time profiling on the left-hand side of Fig. 3.1, it is possible to note that the texture coding represents only 13.7% of the encoder computational effort, whereas depth maps coding represents 86.3% when considering the AI scenario. Detailed profiling of depth coding time is also presented in Fig. 3.1, where DMM-1 is the most time-consuming encoding tool being responsible for 27.2% of the depth maps intra-computational effort, followed by residual encoding tools with 26.6 and 25.3% for TQ (Transform/Quantization) and SDC, respectively. The "Intra HEVC" considers the evaluation of the prediction modes inherited from HEVC texture coding and represents 12.1% of the time spent in depth maps intra coding. Together, the remaining encoding modes, DMM-4 and DIS, represent less than 9%. Even though these modes do not represent high encoding time when compared to other tools, they can provide significant improvements in 3D-HEVC encoding efficiency (Sanchez et al. 2018b).

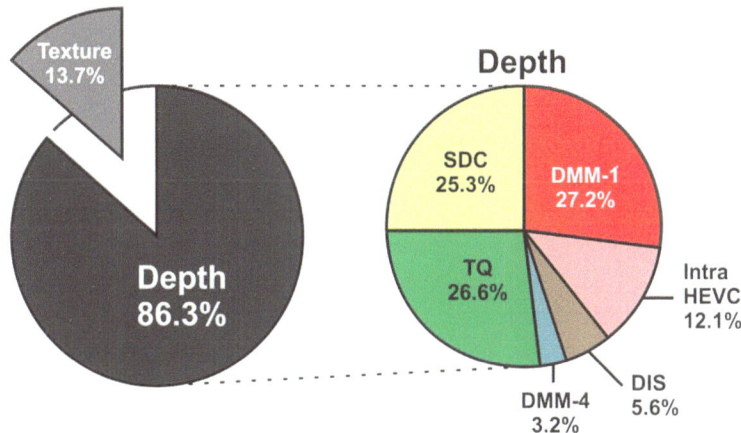

Fig. 3.1 Computational effort distribution in 3D-HEVC intra-frame prediction

These results can be explained since depth maps are composed of large homogenous regions and the DIS typically fits better with these kinds of regions; however, depth maps also present edge and gradient regions, where DMMs and Intra HEVC are crucial to provide high encoding efficiency.

Therefore, according to the data presented in Fig. 3.1, one can conclude that depth maps intra coding is a costly computational task for the 3D-HEVC encoder. The highest computational effort is mainly concentrated in DMM-1 and residual encoding flows. In order to reduce this computational effort, some works proposed to avoid the entire DMM-1 evaluation (Sanchez et al. 2015a) and decrease the number of modes evaluated in residual encoding flow (Sanchez et al. 2017a).

However, when hardware architectures able to achieve real-time processing with low power dissipation are considered, new challenges can be noticed. For instance, in the residual coding flow, limited parallelism can be exploited since these steps share data with entropy encoding that works with a sequential approach. Besides, in DMM-1 the number of wedgelets to be evaluated depends on the block size being predicted (1,908 in total—see Sect. 2.1.4.1) and these wedgelets must be stored in memory to be evaluated.

Many wedgelets patterns for both DMM-1 encoding and decoding processes are allowed. This storage of all wedgelets patterns is undesirable, mainly considering 3D-HEVC systems implemented in dedicated hardware. As discussed before, DMM-1 requires 183,264 bits to store all wedgelets patterns (Zhang et al. 2014; Ikai et al. 2015) of all block sizes supported by the encoding tool (i.e., square-shaped block sizes ranging from 4×4 to 32×32), according to the 3D-HTM implementation. Also, the DMM-1 algorithm requires continuous access to the wedgelets patterns on memory and intensive processing, which demands high energy consumption. Thus, to obtain performance and energy consumption compatible with real-time processing of 3D-HEVC high-definition videos, it is necessary to reduce memory size, memory accesses and processing effort related to DMM-1 calculations.

3.1.2 3D-HEVC Inter-Frames and Inter-View Predictions

ME and DE are among the costly steps of video encoders regarding energy and memory (Afonso et al. 2019a), as previously discussed. Considering the similar behavior of ME and DE in the Inter-frames and Inter-view predictions, these two encoding tools will be analyzed together in this section.

Both ME and DE encoding tools are evaluated in the encoder at the level of PU (Prediction Units) blocks, and the 3D-HEVC allows ME/DE to be performed with 24 block sizes, ranging from 4×8 up to 64×64 pixels in order to improve coding efficiency. In this scenario, the selection of the best match is a costly evaluation process, which is required to achieve the best tradeoff between compression and distortion. This way, ME/DE demands a huge number of block-matching operations leading to intense processing, memory communication and, consequently, high energy consumption.

Memory access is one of the main components regarding energy consumption and performance considering ME/DE hardware designs due to the number of candidate blocks that the BMA compares (Zatt et al. 2011b). Data-reuse schemes such as Level-C (Chen et al. 2006) and other dedicated solutions (Zatt et al. 2011a, b; Sampaio et al. 2013) have been used in ME and ME/DE to reduce the memory bandwidth, as previously presented.

The work (Afonso et al. 2019a) presents an evaluation of the external memory traffic considering 3D-HEVC encoding of three-view (three texture pictures plus their respective depth maps) HD (High Definition) 1080p MVD videos using HOTZS (Hardware-Oriented TZS) static scheduling and HDS (Horizontal Disparity Search) fast block-matching algorithm (this book details these algorithms in Sect. 5.1). In this work, search patterns that were much less complex than the original TZS algorithm were adopted and these algorithms considered only 32×32 and 16×16 block sizes. When no on-chip SRAM is employed, i.e., the ME/DE processing unit communicates directly to the external memory, the demanded communication for real-time encoding would reach 92 GB/s, as demonstrated in Fig. 3.2. Given that current memory technologies, such as the embedded low-power memories, provide no more than 25.6 GB/s in ideal cases (Micron 2014), this target performance is unfeasible. Considering the use of an on-chip SRAM to prefetch and store the SW used for the block-matching (360kB), the communication is reduced to 9.3 GB/s. When the Level-C scheme (see Sect. 2.1.2.2.1) is adopted in addition to the on-chip SRAM memory, the memory bandwidth drops to 3.1 GB/s, leading to a total communication reduction of 96%, as depicted in Fig. 3.2. Nevertheless, the energy consumption related to the external memory communication, and on-chip SRAM dynamic/static consumption is still high, and it hinders the implementation of 3D real-time systems, mainly focusing on handheld devices.

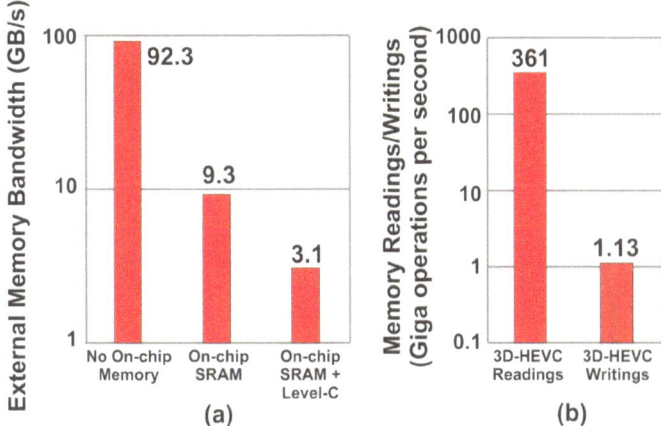

Fig. 3.2 **a** ME/DE external memory bandwidth for three memory organizations (without on-chip memory, with on-chip SRAM and with on-chip SRAM using the Level-C data reuse scheme); and **b** ME/DE memory reading and writing for 3D-HEVC

Regarding memory readings and writings, the work (Afonso et al. 2019a) estimated the number of DE/ME memory operations considering the 3D-HEVC encoding of HD 1080p MVD videos with the 24 possible PU sizes (full RDO cost). The number of encoded views was defined in order to attempt a system capable of delivering nine views after the decoding process. 3D-HEVC requires 1.13×10^9 memory writings and 361×10^9 memory readings per second to encode three HD 1080p views (texture plus depth maps).

Although feasible, the energy consumption related to memory hierarchy remains high due to external memory communication and on-chip SRAM dynamic and static consumption. Thus, a memory hierarchy featuring on-chip SRAM storage is mandatory. However, there is still a need to reduce external memory communication further, reducing on-chip SRAM size (to reduce static/leakage consumption), and managing the memory hierarchy.

Computational effort analyses considering 3D-HEVC ME/DE steps are incipient in the literature. Actually, some works, such as (Afonso et al. 2016), evaluated ME computational effort considering a 2D approach with the HEVC Reference Software (HM—HEVC Test Model) using the Common Test Conditions for HEVC standard (2D videos; Bossen 2013), but the presented results cannot be extrapolated to the ME/DE behavior in a 3D-HEVC context along with MVD format. Therefore, this chapter presents an evaluation of the computational effort related to the ME and DE steps as well as the relation between ME, DE, and Intra-frame prediction and the other tools used in the 3D-HTM. The experiments considered the RA (Random-Access) configuration (Bossen 2013), and two of the video sequences (*Balloons* and *Undo_Dancer*) defined in CTC (Müller et al. 2014) with two QP (Quantization Parameter) values (30 and 39) also defined in CTC. It is important to emphasize that, when considering the RA configuration, Inter-frames prediction,

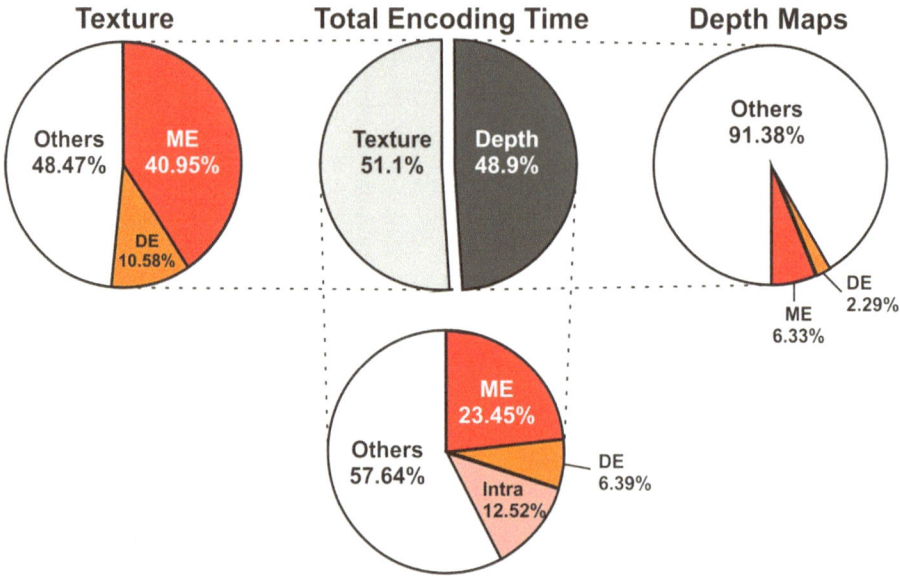

Fig. 3.3 3D-HEVC computational effort distribution when running in RA configuration

Inter-view prediction, and Intra-frame prediction tools are available to encode texture and depth maps.

As one can notice in Fig. 3.3, the time spent with each one of the channels (texture and depth maps) is similar considering an encoder configuration that uses the ME/DE steps, so texture used 51.1% of the encoding time, whereas depth maps used 48.9%.

Regarding the texture channel, a significant portion of the computational effort is due to the ME/DE steps, where ME is responsible for 40.95% of the spent time, DE uses 10.58% of encoding effort, and the other encoding tools (Intra HEVC, Transform, Quantization, etc.) use the remaining 48.47% of the time, as depicted in Fig. 3.3. Considering the depth maps, the portion of computational effort due to the ME/DE is much smaller, where ME uses 6.33% of the time, DE uses 2.29% of the time, and the other encoding tools use 91.38% of the encoding time (Intra HEVC, DMM-1, DMM-4, DIS, Transform, Quantization, SDC, etc.). This behavior is expected since Intra-frame prediction tools are available in the RA configuration and the depth maps tend to be more efficiently encoded when using the novel Intra-frame prediction tools designed for this type of information, as discussed in the previous subsection (Sect. 3.1.1).

Figure 3.3 also presents the total encoding time related to ME/DE and Intra-frame prediction. Note that since ME/DE steps increase the complexity of the texture encoding and the novel 3D-HEVC intra encoding tools increase the complexity of the depth-map encoding, the time spent with predictions is significant considering all the encoding processes. ME/DE predictions spent 29.84% of the time, while Intra-frame prediction spent 12.52% of the time. The remaining encoding tools spent 57.64% of the encoding time.

The memory and computational effort results presented and discussed in this chapter for both the novel depth 3D-HEVC intra-prediction tools and the 3D-HEVC ME/DE steps evidenced the need for the development of complexity-reduction strategies and VLSI designs focusing on both 3D-HEVC Intra-frame prediction, and Inter-frames and Inter-view predictions.

3.2   3D-HTM Encoding-Tool Analyses

This Section presents a wide encoding-tool analysis through the 3D-HTM reference software, focusing on both the CCO and FCO configurations employed in the 3D-HEVC extension. The results for each 3D-HEVC coding-order approach are presented in the following, and they are also divided into subsections, one related to the Inter-frames and Inter-view predictions and another to the Intra-frame prediction. All results consider the experimental setup presented in Appendix A and the RA temporal configuration. The encoding tools are evaluated from the percentage of pixels encoded with them, called of "Representativeness" in this book. It is important to notice that each texture picture has an associated depth-map and, therefore, the same number of pixels is encoded by each channel considering a complete video. The difference is in the fact that depth maps have only luminance samples while texture pictures have three different samples: luminance, chrominance blue, and chrominance red.

3.2.1   3D-HTM Encoding-Tool Analyses Based on the CCO Configuration

As previously explained, the Conventional Coding Order (CCO) configuration adopts the encoding of the texture pictures before their associated depth maps for all views that compose the 3D-video sequences. Figure 3.4 shows the selection of Inter (Inter-frames and Inter-view predictions) and Intra encoding tools according to the PUs for both texture pictures and depth maps. Note that the number of PUs is different for each channel because the depth maps tend to use bigger PU sizes than the texture pictures. This way, the texture channel uses much more PUs to be encoded. Texture encodes 69.6% of the PUs, whereas depth maps use 30.4% of the PUs. Therefore, analyzing only the selection of the PUs according to the encoding tools can cause some imprecision in the assessment. For this reason, the percentage of pixels encoded with the encoding tools, here called of "Representativeness", was prioritized for the 3D-HTM Encoding-tool analyses.

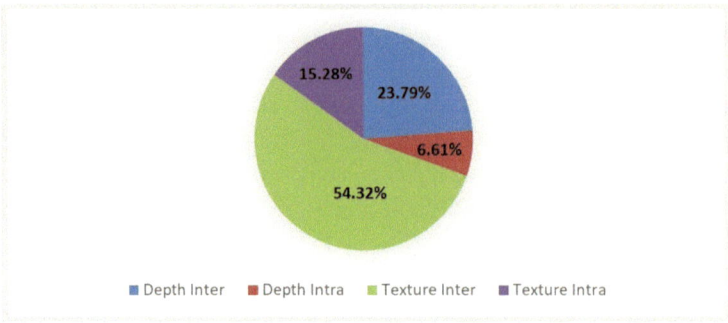

Fig. 3.4 Inter prediction (Inter-frames and Inter-view), and intra-frame PU selection with the CCO approach

The results obtained with the CCO approach were divided according to the type of prediction, Intra-frame or Inter prediction. As presented in Fig. 3.5, the representativeness of the Inter prediction (Inter-frames and Inter-view predictions) is predominant for both the texture and the depth-map pictures. Considering the texture frames (Fig. 3.5a), 97.54% of the pixels are encoded by Inter prediction, while Intra-frame prediction is responsible for encoding only 2.46% of the pixels. In the depth maps (Fig. 3.5c), 92.76% of the pixels are encoded with Inter prediction, and the Intra-frame prediction encodes 7.24% of them. It is important to note that intra-encoded frames are applied to encode the first frame and a new frame at approximately every second to guarantee the references during the encoding process of the other frames. Furthermore, Intra-frame prediction improves the compression gains in some specific situations, such as the coding of object edges. The coding of sharp edges is even more important in an MVD context, where high-quality depth-map coding is needed to efficiently synthesize the intermediate texture views.

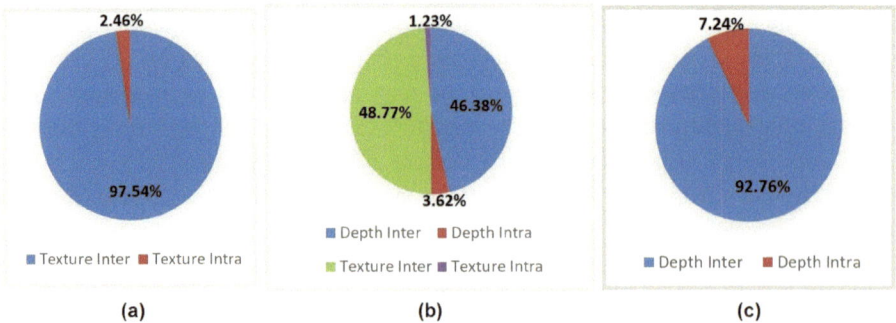

Fig. 3.5 Inter and intra-frame representativeness with a CCO approach: **a** Texture only; **b** both texture and depth; and **c** depth-map only

Regarding encoding time, Inter prediction is well-known as the most costly step of the encoders (Zatt et al. 2013). However, in an MVD context, the Intra-frame prediction is also responsible for an important part of the complexity. The work (Sanchez et al. 2018b) states that 34.51% of the depth-map encoding time is due to the Intra-frame prediction in 3D-HEVC, which surpasses the Inter prediction considering only depth maps (about 19.74%). This way, complexity reduction strategies are important for both the Inter and the Intra-frame predictions of both texture and depth-map coding considering the 3D-HEVC extension. The next subsections present the results related to the Intra-frame and Inter-prediction encoding tools separately.

3.2.1.1 Intra-Frame Prediction

This subsection presents in detail the results obtained from the experiments considering only the Intra-frame prediction and its respective modes. Since the texture and the depth-map pictures present specific characteristics, and the Intra-frame prediction of depth-maps introduces other prediction modes besides the ones used in the texture pictures, the experimental results for the Intra-frame prediction were divided into Texture and Depth-maps to better explore their characteristics and allow the identification of efficient complexity-reduction strategies. The results related to the Texture are shown and discussed in the next subsection. Afterward, the data related to the depth maps are presented.

Texture

According to the performed experiments, 93.73% of the pixels are encoded with 2N × 2N partitions when only Intra-frame prediction and texture pictures are considered, as presented in Fig. 3.6a. The N × N partitions are responsible for encoding the other 6.27% of the pixels (used only in the fourth quad-tree level). Figure 3.6b also shows these data according to the quad-tree level at which these partitions occurred. Considering the results, no CU-depth level presented negligible results, encoding between 19.69 and 34.10% of the Intra-predicted texture pixels. Therefore, Intra-frame prediction simplifications, such as disabling a given partition and/or a quad-tree level, tend to present important impacts when applied to texture pictures, i.e., compression and image-quality losses.

Other data arrangements were used to verify the number of Intra-predicted pixels that were encoded with all 35 modes supported in the texture coding (Planar, DC, and Angular modes). Figure 3.7 shows the percentage of pixels related to each one of these modes. The Planar mode (mode 0 according to the 3D-HTM) is the most important mode according to the results since it encodes 24.03% of the pixels, followed by the DC mode (mode 1), Vertical mode (mode 26), and Horizontal mode (mode 10) with 17.43, 11.07, and 5.92%, respectively. Any of the other modes encode up to 3.11% of the pixels. Also, the four modes that encode more pixels account for 58.45% of the pixels, which is an important result in a scenario with 35 modes. The next subsection presents the Intra-frame prediction results considering depth maps.

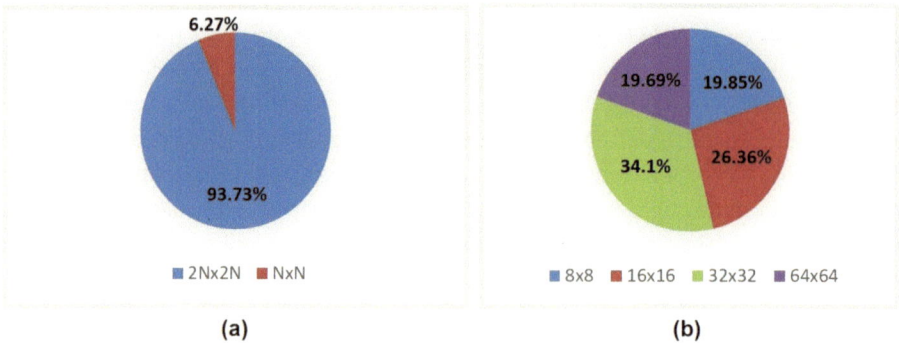

Fig. 3.6 Intra-frame prediction representativeness with a CCO approach considering texture pictures: **a** According to the partition type; **b** according to the CU-depth level

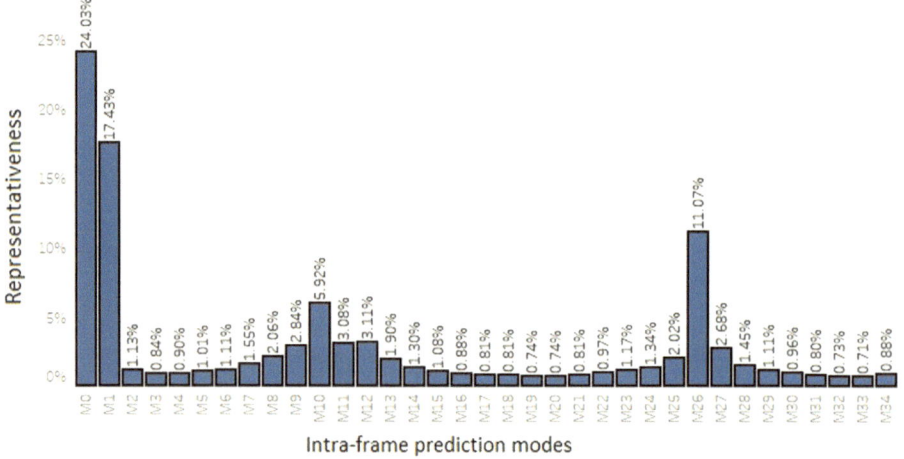

Fig. 3.7 Intra-frame prediction representativeness considering texture pictures according to the encoding mode with a CCO approach

Depth Maps

This subsection presents and discusses the results that correspond to the 3D-HEVC Intra-frame prediction and depth-map pictures.

In the case of the depth-map coding with Intra-frame prediction, it is necessary to analyze the DIS mode first. As previously explained, the DIS mode was introduced in 3D-HEVC to encode depth-maps, and it tends to present better results in this type of picture. Figure 3.8a separately shows the percentage of pixels encoded with the DIS mode and the other Intra-frame tools for depth maps. It is important to notice that the DIS mode is widely used, encoding 81.13% of the Intra-predicted depth-map pixels (used in 2N $\times$ 2N

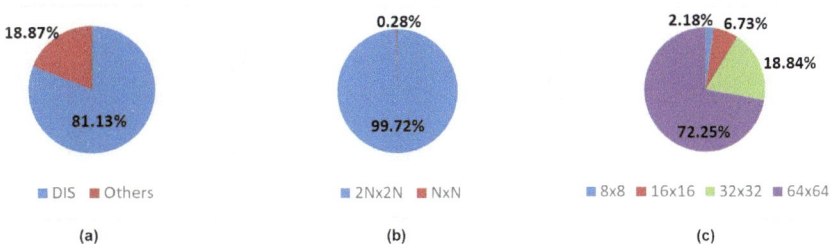

Fig. 3.8 Intra-frame representativeness with a CCO approach for depth maps: **a** Using DIS tool; **b** according to the partition type; **c** according to the CU-depth level

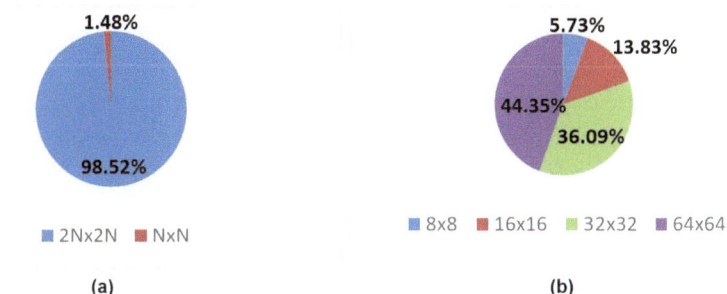

Fig. 3.9 Intra-frame prediction representativeness with a CCO approach for depth maps (except DIS) according to: **a** Partition mode; **b** CU-depth level

partitions only because it is used at the CU level). Considering the remaining pixels and modes, 18.87% of the pixels are encoded and distributed for 37 Intra modes for depth maps (35 HEVC modes and two depth modeling modes).

To better evaluate the Intra-predicted depth-map results and, given the wide usage of the DIS mode, the following results disregard the DIS mode.

Even disregarding the DIS mode by considering only the other modes, the N × N partitions present an insignificant impact in the coding of depth maps, accounting for about 1.48% of the pixels, as presented in Fig. 3.9a. However, considering the CU depth selection, the DIS mode pushes up the 64 × 64 CU-depth level representativeness when compared to the remaining CU depths, as one can observe comparing Figs. 3.8c and 5.8b.

Figure 3.9b shows the representativeness of the Intra-predicted depth maps according to the CU-depth level. As presented in Fig. 3.9b the depth-maps behavior is so different when compared with the CCO Intra-predicted pixels considering texture pictures. While in the CCO texture pictures, the Intra-predicted blocks are usually encoded with all quad-tree levels almost equally distributed, in the depth maps, the encoding with smaller block

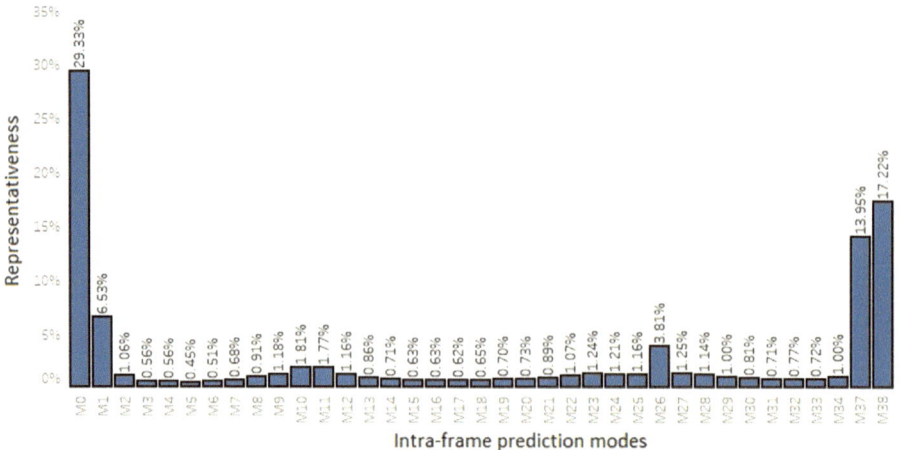

Fig. 3.10 Intra-frame prediction representativeness considering depth-maps according to the encoding mode (except DIS) with a CCO approach

partitions is less frequent than the encoding with bigger partitions. Furthermore, considering the DIS mode, this predominance is even higher, as previously presented (Fig. 3.8c). Nevertheless, conclusions about the importance order of the CU-depth levels in depth maps must be carefully drawn due to the different aspects of the depth maps in which borders must be preserved.

Also, the depth-map modes that are more important in the encoding process were evaluated. As the DIS mode is the most representative Intra-frame depth-map mode, this mode was disregarded in this analysis. Figure 3.10 shows the representativeness of the Intra-frame depth-map modes at which the four most representative modes are Planar (mode 0), DMM-4 (mode 38), DMM-1 (mode 37), and DC (mode 1), with 29.33, 17.22, 13.95, and 6.53% of representativeness, respectively. Together, these four modes are responsible for 67.03% of the Intra-predicted depth-map pixels disregarding the DIS mode. It is important to note that no other mode encodes more than 3.81% of the pixels, and 37 modes can be employed in the depth-maps disregarding the DIS mode (the same texture modes and two DMM modes). Also, when these four most representative modes and the DIS mode are considered, they are responsible for 93.78% of the encoded pixels. By including the six more representative modes, which include the Horizontal (mode 10) and Vertical (mode 26) modes, the percentage of encoded pixels increases to 72.65% disregarding the DIS mode and increases to 94.84% when considering the DIS mode. The DMM modes (modes 37 and 38) were introduced in the 3D-HEVC, especially to deal with depth-map pictures, and they present important results in compression. However, these modes are associated with a large complexity, so they are not applied in the first CU-depth level (64×64 CUs).

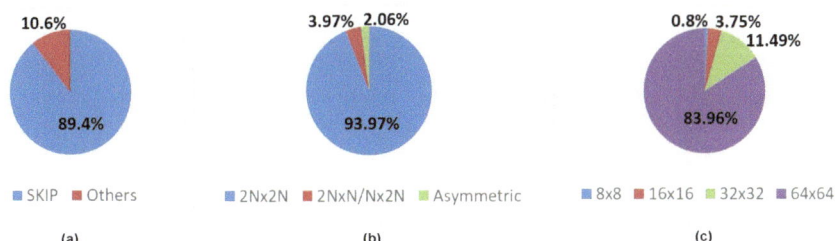

Fig. 3.11 CCO Inter prediction representativeness considering texture pictures: **a** Using skip mode; **b** according to the partition types; **c** according to the CU-depth level

3.2.1.2 Inter-Frames and Inter-View Predictions

This subsection presents the results focusing on the CCO approach and the Inter-frames and Inter-view predictions. As previously mentioned, given the different aspects related to the Texture and Depth Maps, the encoding process of each type of picture adopts different choices along the encoding process. This way, the results are also presented separately in two subsections to Inter prediction, the first considering texture pictures, and the second, depth maps.

Texture

As previously mentioned, about 95.15% of the pixels are encoded using the Inter Prediction (Inter-frames and Inter-view predictions), as presented in Fig. 3.5. Considering only CCO Inter-predicted pixels and the texture pictures, the percentage of pixels is even higher, about 97.54%. As previously explained, the Inter prediction presents several modes and partitions. However, this expressive representativeness of the Inter prediction is mainly due to the Skip mode, as presented in Fig. 3.11a. The percentage of Inter-predicted texture pixels considering the Skip mode is 89.40%, which is applied to any CU-depth level of $2N \times 2N$ partitions, dealing with the $2N \times 2N$ partition to reach 93.97% of the pixels, as presented in Fig. 3.11b. The representativeness of the Inter prediction according to the CU-depth level can be observed in Fig. 3.11c. When the results are evaluated considering the quad-tree level of the CUs, it is possible to observe that the first CU-depth level is responsible for encoding 83.96% of the CCO Inter-predicted texture pixels. Considering the two initial quad-tree levels, the percentage of pixels is 95.44% and, if the three first quad-tree levels are considered, this percentage is 99.2%. This way, the last quad-tree level (8×8 CUs) is responsible for only 0.8% of the encoded pixels. The 64×64 CU-depth level corresponds to the major part of the pixels pushed up by the Skip mode, as well as the $2N \times 2N$ partition.

By analyzing the results regarding the selected partitions and CU-depth levels, the huge representativeness of the Skip mode deals with the $2N \times 2N$ partition and 64×64 CU-depth level to encode the major part of the Inter-predicted texture pixels, as previously discussed. This way, the data were also organized, disregarding the Skip and detailing the

Fig. 3.12 CCO Inter prediction representativeness considering texture pictures (except Skip): **a** According to the partition type; **b** according to the CU-depth level

partitions and CU-depth levels selected with the other modes, as presented in Fig. 3.12. With this data arrangement shown in Fig. 3.12, it is possible to verify that the DBBP, Merge, and ME/DE encoding tools have considerable use of the 2N × N and N × 2N partitions. Despite these partitions being less representative than 2N × 2N partitions considering the CCO Inter prediction of texture pictures, they can present important impacts on the image quality and compression. It is important to notice that 2N × 2N, 2N × N, and N × 2N partitions are responsible for 97.94% of the total CCO Inter-predicted texture pixels (or 80.65% disregarding Skip mode—see Fig. 3.12a). However, the importance of the smaller CU-depth levels is notable, disregarding the Skip mode since the 8 × 8 and 16 × 16 CU depths are responsible for encoding 17.01% of the pixels, even covering a smaller area per block, as depicted in Fig. 3.12b.

In Fig. 3.13, the data are presented disregarding the Skip mode but detailing the other encoding tools. It is possible to notice that Merge and Disparity/Motion Estimations present a similar behavior, whereas they encode a significant area when compared to the DBBP mode. Whereas the Merge and ME/DE encodes 47.17 and 48.49% of pixels, respectively, the DBBP is responsible for encoding 4.34% of the pixels. Note that the DBBP mode is only applied to the 2N × N and N × 2N PU partitions, and to the 64 × 64, 32 × 32, and 16 × 16 CU depths. The next subsection focuses on the results of the CCO Inter prediction considering depth-map pictures.

Depth Maps

Considering the depth-map encoding process, the CCO Inter prediction corresponds to 92.77% of the encoded pixels. Similarly, with the Inter prediction of the texture, this important value is mainly due to the Skip mode. In the depth maps, 96.72% of the CCO Inter-predicted pixels used Skip mode, as presented in Fig. 3.14a. Note that the Skip mode can be applied to any quad-tree level of partitions 2N × 2N.

Regarding partition types, the 2N × 2N partition is responsible for encoding 99.68% of the CCO Inter-predicted depth-map pixels, as presented in Fig. 3.14b. By analyzing the data regarding the quad-tree level at which the blocks were encoded in Fig. 3.14c, it

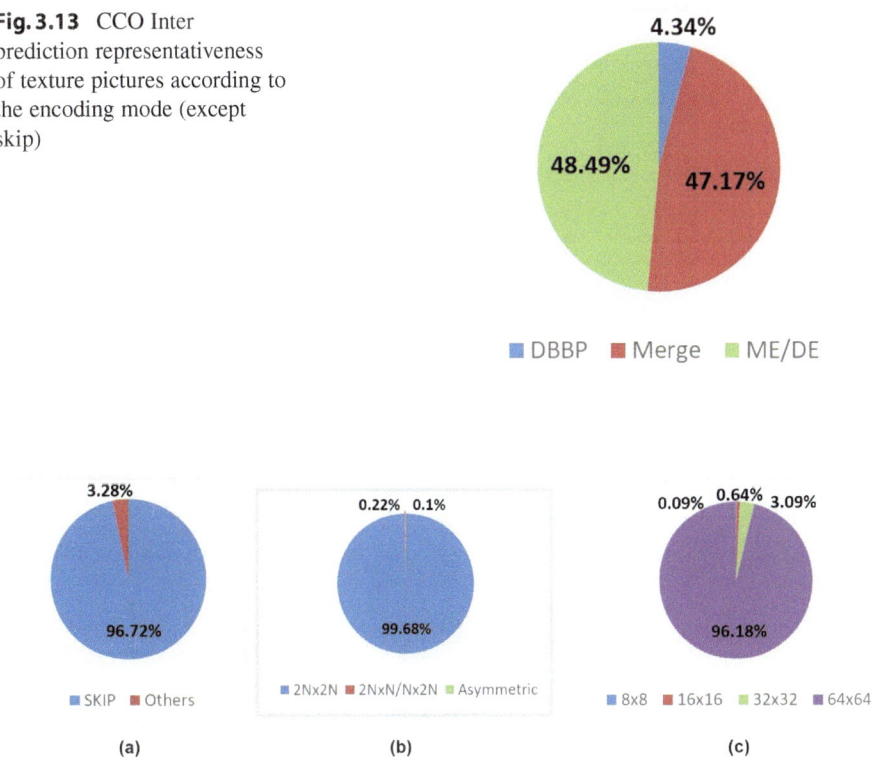

Fig. 3.13 CCO Inter prediction representativeness of texture pictures according to the encoding mode (except skip)

4.34%

48.49%

47.17%

■ DBBP ■ Merge ■ ME/DE

3.28%

96.72%

■ SKIP ■ Others

(a)

0.22% 0.1%

99.68%

■ 2Nx2N ■ 2NxN/Nx2N ■ Asymmetric

(b)

0.09% 0.64% 3.09%

96.18%

■ 8x8 ■ 16x16 ■ 32x32 ■ 64x64

(c)

Fig. 3.14 CCO Inter prediction representativeness considering depth maps: **a** Using skip mode; **b** according to the partition type; **c** according to the CU-depth level

is possible to conclude that the absolute majority of pixels are encoded with the initial quad-tree levels. The 64 × 64 and 32 × 32 CU depths are responsible for encoding 99.27% of the pixels. In depth maps, this predominance of areas encoded with 2N × 2N partitions and CU-depth levels with larger CU sizes tends to be pushed up by the Skip mode, as well as in the texture pictures. This way, an evaluation disregarding the Skip mode is presented in the following.

A data arrangement that disregards the Skip mode is presented in Fig. 3.15. This way, the predominance of the 2N × 2N partitions can be really verified for the other encoding tools. Even disregarding the Skip mode, the 2N × 2N partition encodes 90.48% of the CCO Inter-predicted depth-map pixels. Similarly, even disregarding the Skip mode, the 64 × 64, and 32 × 32 CU depths are responsible for encoding 95.15% of the pixels. In the sequence, a further analysis according to the encoding mode is presented.

The results disregarding the Skip mode are also separated according to the encoding tools, and they are presented in Fig. 3.16. Over again, it is possible to observe that the tool behaviors for depth maps are similar with and without the Skip mode. Merge mode

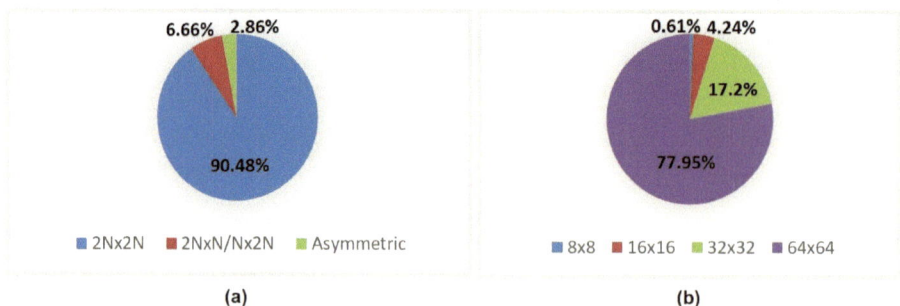

Fig. 3.15 CCO Inter prediction representativeness considering depth maps (except Skip): **a** According to the partition type; **b** according to the CU-depth level

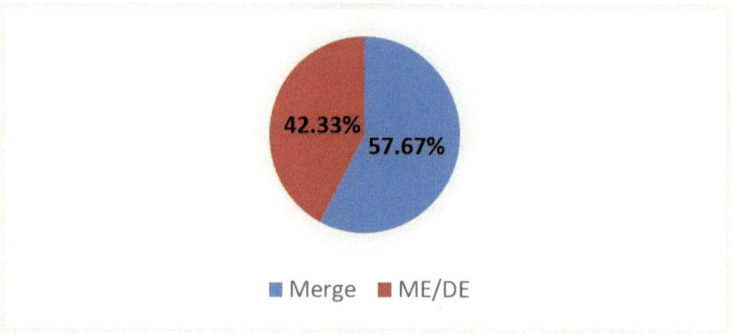

Fig. 3.16 CCO Inter prediction representativeness of depth maps according to the encoding mode (except skip)

encodes more pixels than ME/DE encoding tools, but both Merge, and ME/DE steps are significant in the encoding process.

Block-Size Analysis

As previously mentioned, the major part of the computational effort of the 3D-HEVC is due to the decision of which methods of encoding and PU sizes must be used in the Inter prediction, since until 24 PU sizes can be evaluated during the encoding process if the full RDO cost was considered. All these 24 PU sizes must be processed by other encoding tools (Transforms and Quantization, for instance) to define which size presents the best tradeoff between compression and image quality. The Merge-based encoding, along with Motion and Disparity Estimations, are important steps in the current video coding standards due to the compression rates propitiated by them. To reach such performance, Merge mode, Motion and Disparity Estimations support 24 PU sizes.

In conclusion, this process has a high cost, and a reduction in this computational effort is highly desirable to allow energy-efficient and real-time systems. To allow the identification of computational-effort reduction strategies for the Inter prediction (both Inter-frames and Inter-view predictions), block-size analyses are presented in the following for both texture and depth-maps. These analyses disregarded the Skip mode (less complex mode applied to only four block sizes) and they focused on the more complex steps.

It is possible to infer that a simple way to reduce the computational effort is by reducing the PU sizes that must be compared in the Inter prediction, mainly due to the ME/ DE steps. However, the real impact in terms of compression and image quality of using some specific PU must be evaluated. To support this idea, the incidence of each PU size in the Inter prediction and its representativeness on the frames were investigated. Hence, the 3D-HTM code was modified with the aim of extracting those data.

All the eight sequences defined by the CTC were encoded in the RA temporal configuration and according to the configurations recommended in that document. Figure 3.17 shows the percentage of PU sizes selection in the Inter prediction, on average, considering the texture pictures. The values are presented separately for all 24 PU sizes according to the CU-depth level. Considering only the data referred to the ME/DE and Merge steps (and DBBP mode for the texture), i.e., disregarding Skip mode, the 8×8 PU size is the most frequently selected block size with 14.08% of selections and the second most often selected size is the 16×16 with 10.98% of selections. Note that the other two square-shaped block sizes allowed in the Inter prediction (32×32 and 64×64 PU sizes) are poorly selected when compared to other sizes (seventh and fifteenth more selected sizes only).

The percentage of PU size selection suggests that some PU sizes, such as the 8×8 PU size, have great importance during the coding process. However, since bigger PUs are more representative in the image, evaluating the percentage of pixels that were covered by each PU size is important. Bigger PUs, such as the 16×16, even being less frequent, may cover a larger area and, therefore, they can be more relevant to the coding process.

To further evaluate this hypothesis, the data about the selection of the PU sizes were adjusted considering the image representation of each PU size. This analysis, as depicted in Fig. 3.18, shows that bigger sizes (such as 64×64 and 32×32) are more representative in the video sequences, even being less frequent and even when the skip mode is not considered. Whereas 64×64 PU size is the most representative size, encoding 20.61% of the pixels, 32×32 PUs are the second most representative PUs by encoding 14.2% of the pixels. Note that the square-shaped PU sizes are the most representative sizes at each of the CU depth levels. Figures 3.17 and 3.18 show that square-shaped PUs are both frequent and representative when compared to the non-square-shaped PUs in the texture picture encoding process. Note that the 8×8 PU size is the most frequent and the 16×16 PU size is the second most frequent, whereas the square-shaped sizes 64×64, and 32×32 are the most representative sizes.

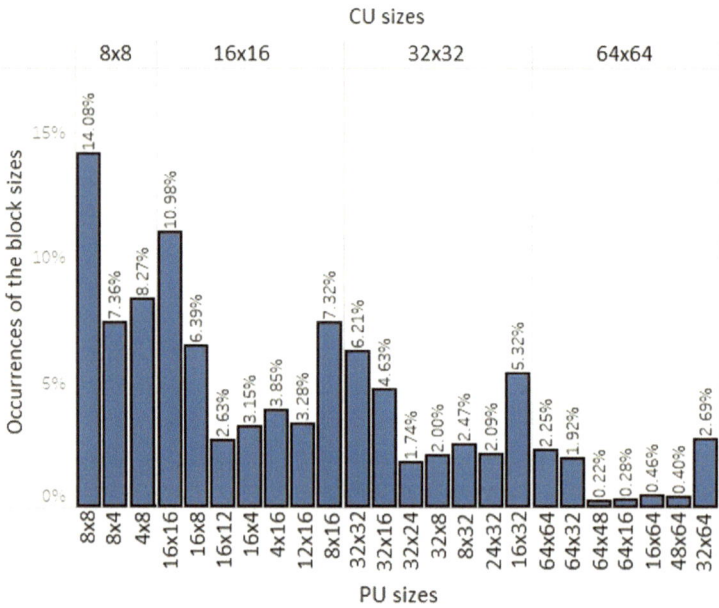

Fig. 3.17 Occurrences of the block sizes in the CCO Inter prediction considering texture pictures (except skip mode)

Figure 3.19 shows the average percentage of PU size selection in the Inter prediction considering the depth maps. The values related to the depth maps are presented separately for all 24 PU sizes according to the CU-depth level. Considering only the data referred to the ME/DE and Merge steps, i.e., disregarding Skip mode, the 64×64 PU size is the most frequently selected block size with 25.09% of selections, followed by the other three square-shaped block sizes allowed in the Inter prediction, 32×32, 16×16, and 8×8 PU sizes. The 32×32 PU size is the second most often selected size with 20.03% of selections, the 16×16 PU size is the third most often selected PU size with 19.47% of selections, and the 8×8 PU size is the fourth most often selected PU size with 11.69% of selections. Note that this behavior is different from that obtained in the texture pictures since bigger PUs are more selected when considering depth maps. Furthermore, it is possible to observe that the four square-shaped block sizes cover 76.28% of the selections related to depth maps, and the other 23.72% are distributed among 20 PU sizes. For depth maps, the data about the PU size selection were also adjusted to consider the representativeness. Figure 3.20 depicts the percentage of pixels that were encoded by each PU size. This analysis shows that the three bigger square-shaped sizes, 64×64, 32×32, and 16×16, are the three most representative sizes, encoding 72.08, 14.38 and 3.5% of the pixels, respectively. The other square-shaped PU size, the 8×8, is the eighth most representative size. Note that the square-shaped PU sizes are the most representative sizes in each one of

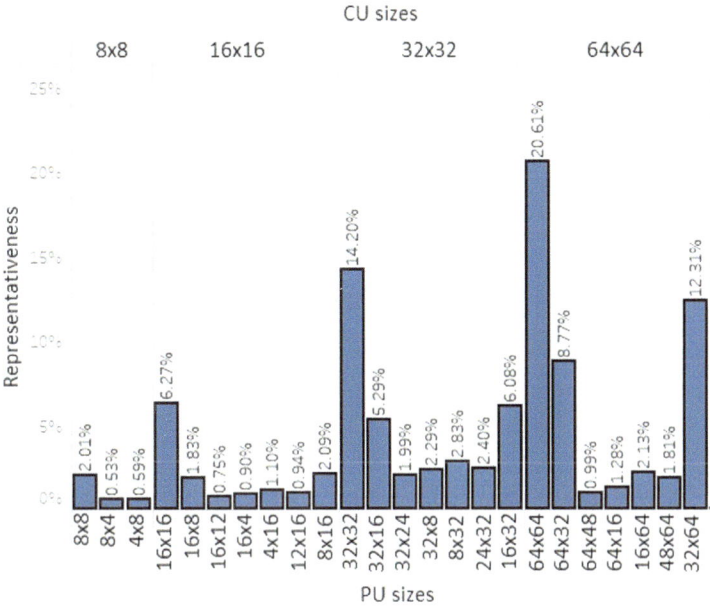

Fig. 3.18 Representativeness of the block sizes in the CCO Inter prediction considering texture pictures (except skip mode)

the CU-depth levels and, together, they cover 90.48% of the encoded pixels considering depth maps. Figures 3.19 and 3.20 showed that square-shaped PUs are both frequent and representative when compared to the non-square-shaped PUs in the depth-map encoding process. Note that the three bigger square-shaped sizes, 64×64, 32×32, and 16×16 PU sizes, are both the three most frequent and the three most representative sizes, whereas the 8×8 PU size is the fourth most frequent.

The 3D-HEVC evaluations allowed verifying the occurrences of the PU sizes that are most selected and most representative during the encoding considering the CCO Inter prediction. From these evaluations, it is possible to note that the square-shaped PU sizes have the two most selected sizes (16×16 and 8×8) and they have the two most representative sizes (64×64 and 32×32) considering the texture pictures. Considering depth maps, it is possible to note that the square-shaped PU sizes are the four most selected sizes (64×64, 32×32, 16×16, and 8×8) and they have the three most representative sizes (64×64, 32×32, and 16×16). Finally, the four square-shaped PU sizes are the most representative sizes in each one of the CU depth-levels for both texture pictures and depth maps. Based on these observations, some scenarios that limit the PU sizes in the Inter prediction were investigated targeting a complexity reduction to support energy-efficient hardware design (see Chap. 5).

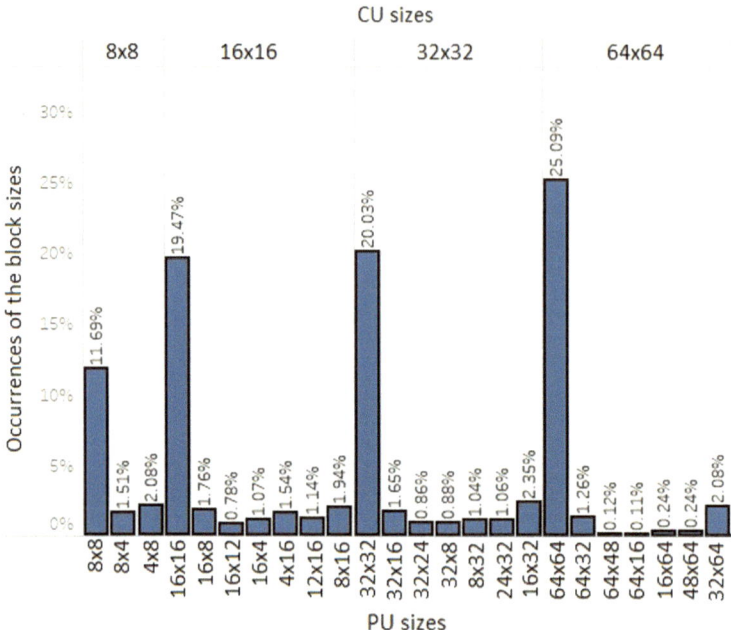

Fig. 3.19 Occurrences of the block sizes in the CCO Inter prediction considering depth maps

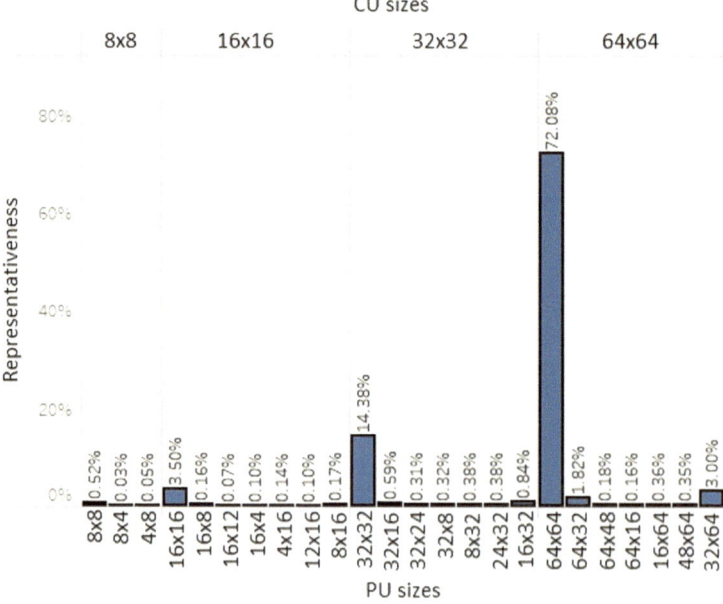

Fig. 3.20 Representativeness of the block sizes in the CCO Inter prediction considering depth maps

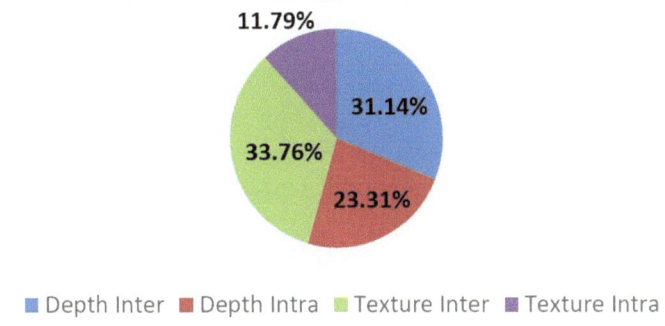

■ Depth Inter ■ Depth Intra ■ Texture Inter ■ Texture Intra

Fig. 3.21 Inter prediction (Inter-frames and Inter-view), and Intra-frame prediction PU selection with the FCO approach

3.2.2 3D-HTM Encoding-Tool Analyses Based on FCO Configuration

As previously explained, the FCO configuration adopts the encoding of the texture pictures before their associated depth maps for BVs and depth maps before texture pictures for all dependent views that compose the 3D-video sequences. Figure 3.21 shows the selection of Inter prediction (Inter-frames and Inter-view), and Intra-frame prediction encoding tools according to the PUs for both texture pictures and depth maps.

Note that the selection behavior is different, considering the FCO configuration, when compared with the CCO approach. In FCO, depth maps encode a higher number of PUs than the texture due to the use of smaller PU sizes. This way, texture encodes 45.55% of the PUs, whereas depth maps use 54.45% of the PUs. In order to obtain a more precise assessment of the encoding-tool behavior, the percentage of pixels encoded with the encoding tools was prioritized for the 3D-HTM encoding-tool analyses in the sequence of the text.

The results obtained with the FCO approach were also divided according to the type of prediction, Intra-frame or Inter prediction. As presented in Fig. 3.22, the representativeness of the Inter Prediction is predominant for both the texture and the depth-map pictures. Considering the texture frames (Fig. 3.22a), 97.62% of the pixels are encoded by Inter prediction, while Intra-frame prediction is responsible for encoding only 2.38% of the pixels. In the depth maps, 82.1% of the pixels are encoded with Inter prediction, and the Intra-frame prediction encodes 17.9% of them (see Fig. 3.22c).

It is important to notice that the the depth intra tools importance is higher in an FCO approach than in a CCO approach, which is expected since dependent views encode the depth maps before texture frames. This way, DIS and DMM-1 modes tend to obtain a better match than Inter prediction tools (especially the Disparity Estimation and Merge

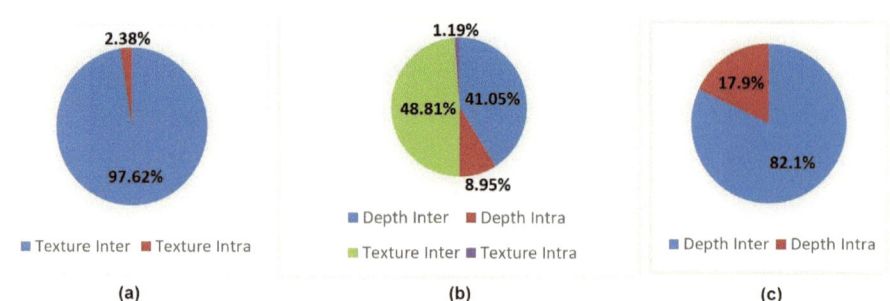

Fig. 3.22 Inter and Intra-frame prediction representativeness with an FCO approach: **a** Texture pictures; **b** texture and depth maps; **c** depth maps

mode), preserving the depth-map characteristics needed to efficiently synthesize the intermediate texture views. The next subsection presents an Intra-frame prediction analysis considering the FCO approach.

3.2.2.1 Intra-Frame Prediction

This subsection presents in detail the results obtained from the experiments considering only the Intra-frame prediction and its respective modes under the FCO configuration. First, the texture encoding process with Intra-frame prediction is exploited, and, in the sequence, the depth-map encoding is analyzed using the experimental results. This way, both texture and depth-map characteristics can be exploited separately to identify efficient complexity-reduction strategies for the Intra-frame prediction in an FCO context.

Texture

According to the developed experiments, 93.74% of the pixels are encoded with 2N × 2N partitions when only Intra-predicted texture pixels are considered using the FCO configuration, as presented in Fig. 3.23a. The N × N partitions are responsible for encoding the other 6.26% of the pixels (used only in the fourth quad-tree level). Figure 3.23b also shows these data according to the CU-depth level at which these partitions occurred. Considering the results, no quad-tree level presented negligible results since the different levels encode between 19.7 and 34.12% of the Intra-predicted texture pixels. In other words, the behavior of the texture picture encoding is very similar for both CCO and FCO approaches.

Other data arrangements were used to verify the number of Intra-predicted pixels that were encoded with all 35 modes supported by the texture coding (Planar, DC, and Angular). Figure 3.24 shows the percentage of pixels encoded with each one of these modes. The Planar mode (mode 0 according to the 3D-HTM) is the most important mode since it encodes 24.30% of the pixels, followed by the DC mode (mode 1), Vertical mode (mode 26), and Horizontal mode (mode 10) with 17.68, 10.35, and 5.96%, respectively. The

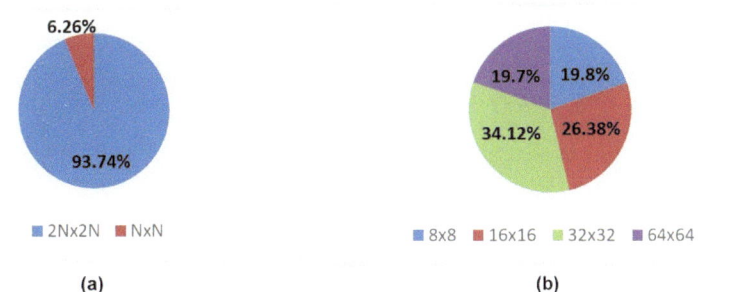

Fig. 3.23 FCO intra-frame prediction representativeness considering texture pictures: **a** According to the partition type; **b** according to the CU-depth level

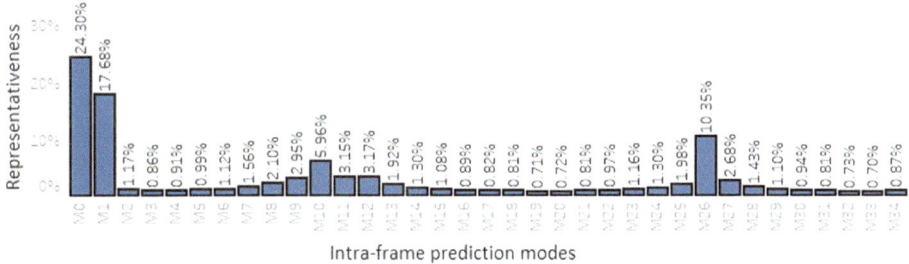

Fig. 3.24 Intra-frame prediction representativeness according to the encoding mode for the texture pictures

other modes encode a maximum of 3.17% of the pixels. Also, the four modes that encode more pixels account for 58.29% of the pixels, which is an important result in a scenario with 35 modes. Note that the behavior of the texture picture encoding related to the mode distribution is also very similar for both CCO and FCO approaches.

The next subsection presents the Intra-frame prediction results considering depth maps.

Depth Maps

This subsection presents and discusses the results that correspond to the Intra-predicted depth-map pixels. As previously explained, in the case of the depth-map coding with Intra-frame prediction, it is necessary to analyze the DIS mode first. Figure 3.25a separately shows the percentage of pixels encoded with the DIS mode and the other Intra-frame tools for depth maps in an FCO approach. It is important to notice that the DIS mode is widely used in the FCO approach as well as in the CCO approach, encoding 86.68% of the Intra-predicted depth-map pixels when using FCO. The 13.32% of remaining pixels

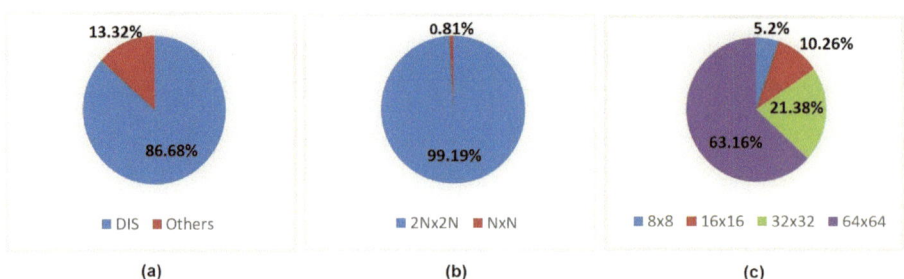

Fig. 3.25 Intra-frame prediction representativeness with an FCO approach for depth maps: **a** Using the DIS mode; **b** according to the partition type; **c** according to the CU-depth level

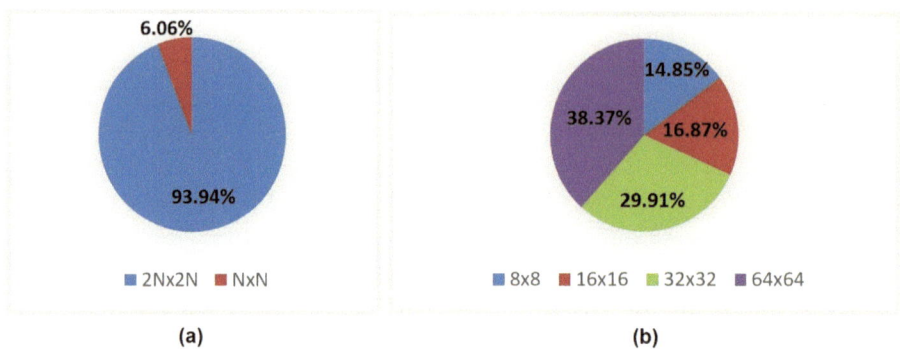

Fig. 3.26 Intra-frame prediction representativeness with an FCO approach for depth maps (except DIS) according to the: **a** Partition mode; **b** CU-depth level

are distributed to be encoded by the 37 Intra modes defined for depth maps (35 HEVC modes and two depth modeling modes).

Even disregarding the DIS mode by considering only the other modes, the N × N partitions present a small impact in the coding of depth maps, accounting for about 6.06% of the pixels, as presented in Fig. 3.26a. However, the N × N partition is not negligible in an FCO approach when compared with a CCO approach (increases from 1.48 to 6.06%) if the DIS encoding tool is disregarded. In terms of CU-depth distribution, the DIS mode pushes up the use of 64 × 64 CU-depth level, as presented in Figs. 3.25c and 3.26b.

To better evaluate the Intra-encoded depth-map representativeness, Figs. 3.26 and 3.27 disregard the DIS mode. Figure 3.26b shows the representativeness of the Intra-predicted depth maps according to the quad-tree level. As presented in Fig. 3.26b, the FCO depth-map encoding has a different behavior when compared with the CCO. While in the CCO approach, the Intra-predicted depth blocks encoded with bigger sizes tend to be more

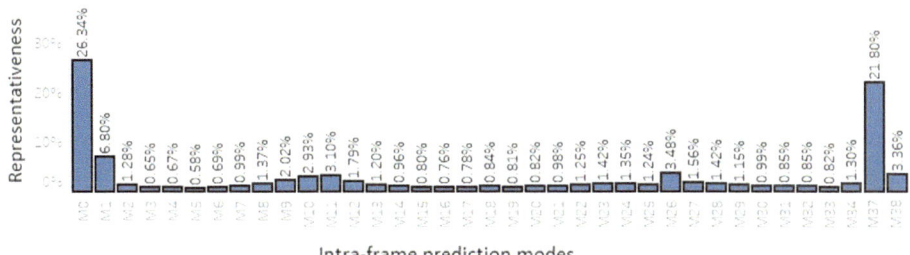

Fig. 3.27 FCO Intra-frame prediction representativeness considering depth-maps according to the encoding mode (except DIS)

representative, in the FCO, the area covered by each CU-depth level is distributed more evenly.

The most important depth-map modes in the encoding process were also evaluated in this subsection. Since the DIS mode is the most representative Intra-frame depth-map mode, this mode was disregarded. Figure 3.27 shows the representativeness of the Intra-frame depth-map modes where the five most representative modes are Planar (mode 0), DMM-1 (mode 37), DC (mode 1), Vertical (mode 26), and DMM-4 (mode 38), with 26.34, 21.80, 6.80, 3.48, and 3.36% of representativeness, respectively. The sixth most representative mode is the Angular 11 with 3.10%, whereas the Horizontal mode (M10) is the seventh most representative mode with 2.93% of representativeness.

Together, these seven modes are responsible for 67.81% of the Intra-predicted depth-map pixels disregarding the DIS mode. It is important to note that no other mode encodes more than 2.02% of the pixels and there are 37 modes that can be employed in the depth-maps disregarding the DIS mode (the same texture modes and two DMM modes).

By including six modes among the seven most representative modes, i.e., Planar, DC, DMM-1, DMM-4, Horizontal, and Vertical modes, the percentage of encoded pixels disregarding the DIS mode is 64.71%, and this number increases to 95.3% when considering the DIS mode. As previously stated, the DMM modes (modes 37 and 38) are associated with a large complexity, so they are not applied in the first quad-tree level (64×64 CU-depth level). The coding with DMM-4 mode covered a reduced area considering the FCO approach, only 3.36% (decreases from 17.22% in CCO to 3.36% in FCO). This behavior is expected since the DMM-4 requires information on texture collocated blocks and, in the FCO approach, this is possible only in the Base View. In dependent views, depth-maps are encoded before texture frames, which disabled the use of DMM-4.

3.2.2.2   Inter-Frames and Inter-View Predictions

This subsection presents the results focusing on the FCO approach and in the Inter Prediction, i.e., focusing on the Inter-view and Inter-frames predictions. The results are also

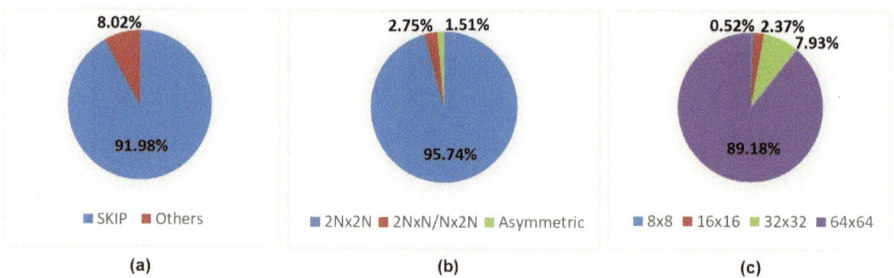

Fig. 3.28 FCO Inter prediction representativeness considering texture pictures: **a** Using the skip mode; **b** according to the partition type; **c** according to the CU-depth level

presented separately in two subsections: one considering texture pictures and another considering depth maps.

Texture

As previously mentioned, most of the pixels are encoded using the Inter prediction in the FCO approach (about 89.86%, see Fig. 3.22). This is also true for FCO Inter-predicted texture pixels, where the percentage of pixels is even higher: 97.62%. As previously explained, this expressive representativeness of the Inter prediction is mainly due to the Skip mode, as presented in Fig. 3.28a. The percentage of Inter-predicted texture pixels considering the Skip mode is 91.98%, which is applied to any quad-tree level of partitions $2N \times 2N$, dealing with the $2N \times 2N$ partition to reach 95.74% of the pixels, as presented in Fig. 3.28b. The representativeness of the Inter prediction according to the CU-depth level can be observed in Fig. 3.28c. When the results are evaluated considering the quad-tree level of the CUs, it is possible to observe that the first quad-tree level is responsible for encoding 89.18% of the FCO Inter-predicted texture pixels. Considering the first and the second quad-tree levels, the percentage of encoded pixels is 97.11% and, if the three first quad-tree levels are considered, this percentage grows to 99.48%. This way, the last quad-tree level (8×8 CUs) is responsible for only 0.52% of the encoded pixels. Both the $2N \times 2N$ partition and the 64×64 CU-depth level representativeness are pushed up by the Skip mode.

This analysis was also done disregarding the Skip and detailing the partitions and CU-depths selected with the other modes, as presented in Fig. 3.29. With this data arrangement shown in Fig. 3.29a, it is possible to verify that the DBBP, Merge, and ME/DE encoding tools are considerably used in the $2N \times N$ and $N \times 2N$ partitions. Despite these partitions being less representative than $2N \times 2N$ partitions considering the FCO Inter prediction of texture pictures, they can present important impacts on image quality and compression. It is important to notice that $2N \times 2N$, $2N \times N$, and $N \times 2N$ partitions are responsible for 98.49% of the total FCO Inter-predicted texture pixels (or 81.03%, disregarding Skip mode). However, the importance of the smaller CU-depth levels is notable, disregarding

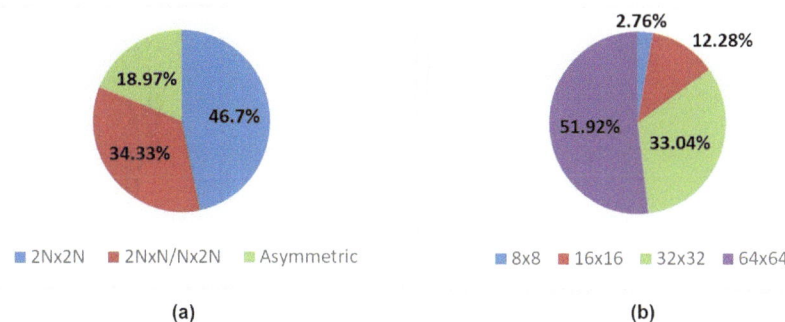

Fig. 3.29 FCO Inter prediction representativeness considering texture pictures (except skip): **a** According to the partition type; **b** according to the CU-depth level

Fig. 3.30 FCO Inter prediction representativeness of texture pictures according to the encoding mode (except skip)

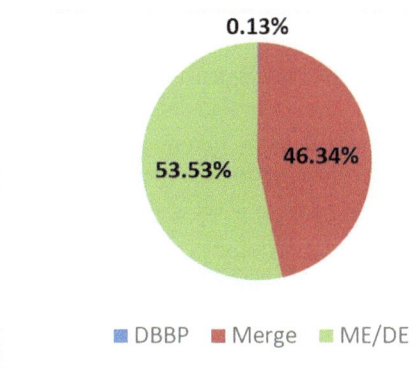

the Skip mode since the 8×8 and 16×16 CU depths are responsible for encoding 15.04% of the pixels, even covering a smaller area per block, as depicted in Fig. 3.29b.

In Fig. 3.30, the data are presented disregarding the Skip mode but detailing the other encoding tools. It is possible to notice that Merge and Disparity/Motion Estimations present a similar behavior, and also they encode a significant number of pixels when compared to the DBBP mode. Whereas the Merge and ME/DE encodes 46.34 and 53.53% of pixels, respectively, the DBBP is responsible for encoding only 0.13% of pixels. Note that the number of pixels encoded using DBBP decreases in an FCO approach when compared to the CCO approach (from 4.34 to 0.13%). The next subsection focuses on the results of the FCO Inter prediction of depth-map.

Depth Maps

Considering the FCO approach and the depth-map encoding process, Inter prediction (Inter-view and Inter-frames) corresponds to 82.1% of the encoded pixels. This higher

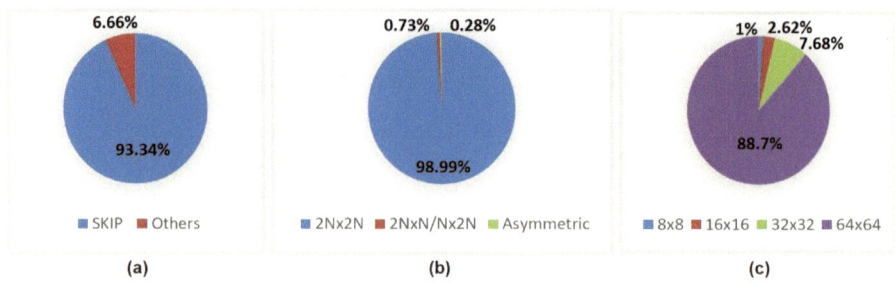

Fig. 3.31 FCO Inter prediction representativeness considering depth maps: **a** Using skip mode; **b** according to the partition type; **c** according to the CU-depth level

value is mainly due to the Skip mode. In the depth maps, 93.34% of the FCO Inter-predicted pixels used Skip mode, as presented in Fig. 3.31a. Regarding partition types, the 2N × 2N partition is responsible for encoding 98.99% of the FCO Inter-predicted depth-map pixels, as presented in Fig. 3.31b. The analysis regarding the quad-tree level at which the blocks were encoded (see Fig. 3.31c) shows that the absolute majority of pixels are encoded with the first quad-tree levels. The 64 × 64 and 32 × 32 CU depths are responsible for encoding 96.38% of the pixels considering depth maps. This predominance of areas encoded with 2N × 2N partitions and CU-depth levels with larger CU sizes tends to be pushed up by the Skip mode, as in the previous cases. This way, again, an evaluation disregarding the Skip mode is presented.

A data arrangement that disregards the Skip mode is presented in Fig. 3.32. This way, the predominance of the 2N × 2N partitions can be confirmed for the other encoding tools. Even disregarding the Skip mode, the 2N × 2N partition encodes 84.8% of the FCO Inter-predicted depth-map pixels. Similarly, even disregarding the Skip mode, the 64 × 64, and 32 × 32 CU-depth levels are responsible for encoding 86.98% of the pixels. In the sequence, a further analysis according to the encoding mode is presented.

The results disregarding the Skip mode and separated according to the encoding tools are presented in .

Figure 3.33 again, it is possible to observe that the tool behavior, for depth maps, is similar with and without the Skip mode. Whereas Merge encodes 48.19% of the pixels, ME/DE Estimations are responsible for encoding 51.81% of the pixels. Both Merge mode and ME/DE steps tend to encode more with 2N × 2N partitions and with the first CU-depth level.

Block-Size Analysis

Figure 3.34 shows the average percentage of PU size selection in the Inter prediction considering the texture pictures and the FCO approach. The values are presented separately for all 24 PU sizes according to the CU-depth level. Considering only the data referred to the ME/DE and Merge steps (and DBBP mode for the texture), i.e., disregarding Skip

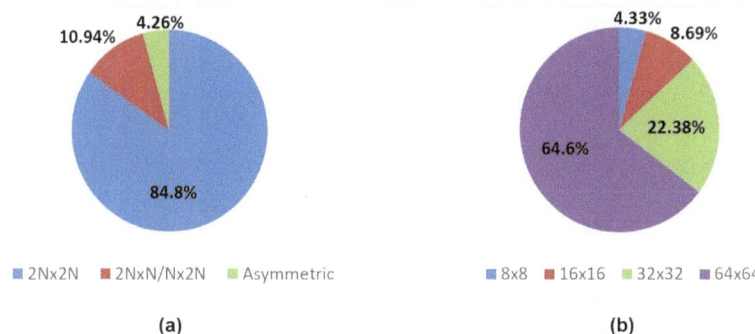

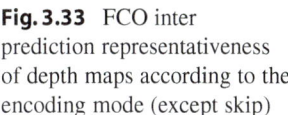

(a)   (b)

Fig. 3.32 FCO inter prediction representativeness considering depth maps (except skip): **a** According to the partition type; **b** according to the CU-depth level

Fig. 3.33 FCO inter
prediction representativeness
of depth maps according to the
encoding mode (except skip)

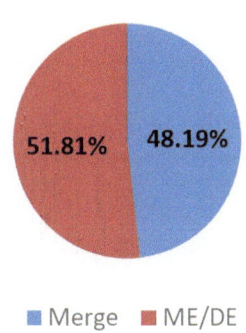

mode, the 8×8 PU size is the most frequently selected block size with 15.03% of selections. The second most often selected size is 16×16, with 11.96% of the selections. Note that the other two square-shaped block sizes allowed in the Inter prediction (32×32 and 64×64 PU sizes) are less selected than the 8×8 and 16×16 sizes. Nevertheless, 32×32 is the fourth most selected size with 6.91%. The 64×64 block size is the thirteenth most selected size, with 3.02% of the PUs.

Since bigger PUs are more representative in the image than the smaller PUs, the percentage of pixels that were covered by each PU size was evaluated. The data about the PU size selection were adjusted considering the percentage of pixels that were encoded by each PU size, considering an average of all tested conditions. This analysis, as depicted in Fig. 3.35, shows that bigger sizes (such as 64×64 and 32×32) are more representative in the video sequences, even being less frequent. Whereas 64×64 PU size is the most

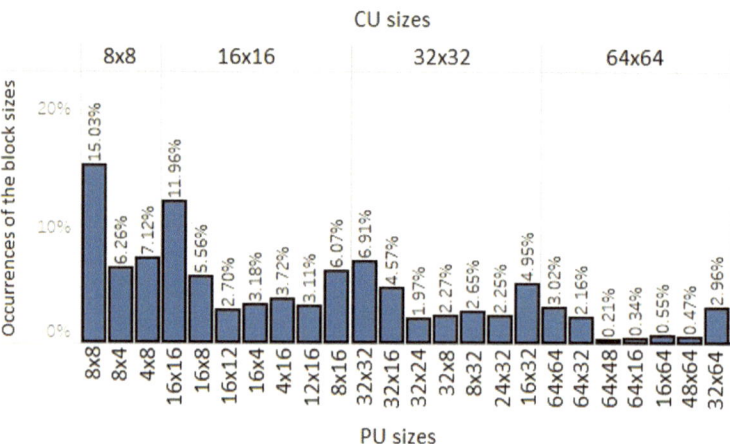

Fig. 3.34 Selection of PU sizes for the Inter prediction considering texture pictures and FCO approach

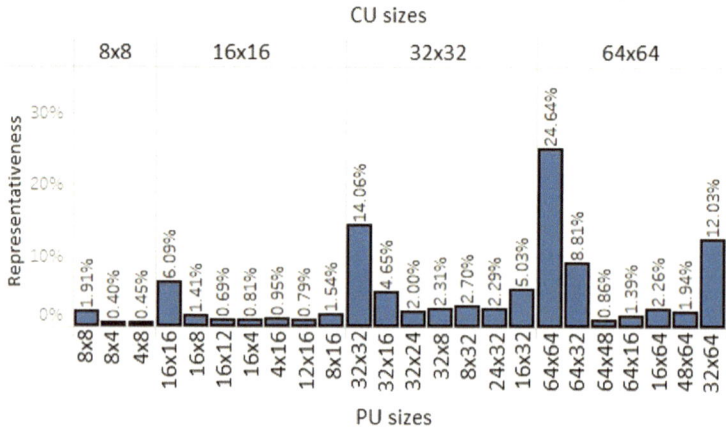

Fig. 3.35 Representativeness of PU sizes for the Inter prediction considering texture pictures and FCO approach

representative size, encoding 24.64% of the pixels, 32×32 PUs are the second most representative PUs by encoding 14.06% of the pixels. Note that the square-shaped PU sizes are the most representative sizes at each of the CU depth levels.

Figures 3.34 and 3.35 show that square-shaped PUs are both frequent and representative when compared to the non-square-shaped PUs in the texture picture encoding process with the FCO approach. Note that the 8×8 PU size is the most frequent, and the 16×16

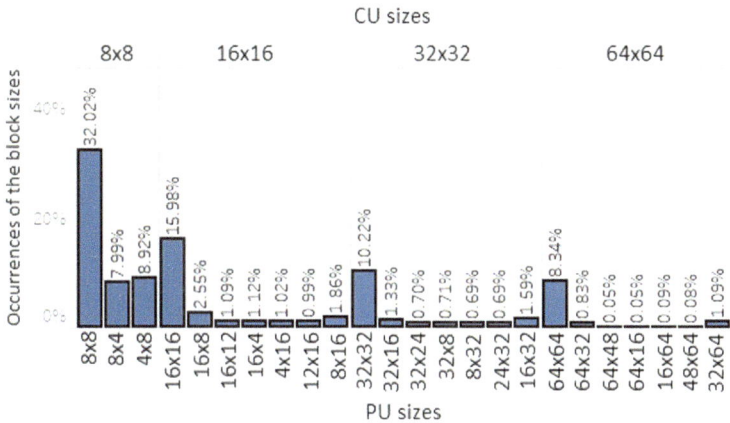

Fig. 3.36 Selection of PU sizes for the inter prediction considering depth maps and FCO approach

PU size is the second most frequent, whereas the square-shaped sizes 64×64, and 32×32 are the most representative sizes.

Figure 3.36 shows the average percentage of PU size selection in the Inter prediction considering the depth maps and the FCO approach. The values related to the depth maps are presented separately for all 24 PU sizes according to the CU-depth level. Considering only the data referred to the ME/DE and Merge steps, i.e., disregarding Skip mode, the 8×8 PU size is the most frequently selected block size with 32.02% of selections, followed by other two square-shaped block sizes, 16×16, and 32×32 PU sizes. The 16×16 PU size is the second most often selected size, with 15.98% of selections, and the 32×32 PU size is the third most often selected PU size, with 10.22% of selections. Another square-shaped block size, the 64×64, is the fifth most often selected PU size, with 8.34% of selections. Note that in an FCO approach, the behavior corresponding to the block sizes is similar for texture pictures and depth maps. The four square-shaped block sizes cover 66.56% of the selections related to depth maps, and the 33.44% are distributed for the other 20 PU sizes.

For depth maps, the data about the PU size selection were also adjusted considering the image representativeness, as presented in Fig. 3.37. This analysis shows that the three bigger sizes (64×64, 32×32, and 16×16) are the most representative sizes by encoding 57.06, 17.49, and 6.84% of the pixels, respectively. The other square-shaped PU size, the 8×8, is the fifth most representative size with 3.43%. Note that the square-shaped PU sizes are the most representative sizes in each one of the CU-depth levels and, together, they cover 84.82% of the encoded pixels considering depth maps. Figures 3.36 and 3.37 show that square-shaped PUs are both frequent and representative when compared to the non-square-shaped PUs in the depth-map encoding process. Note that the three bigger

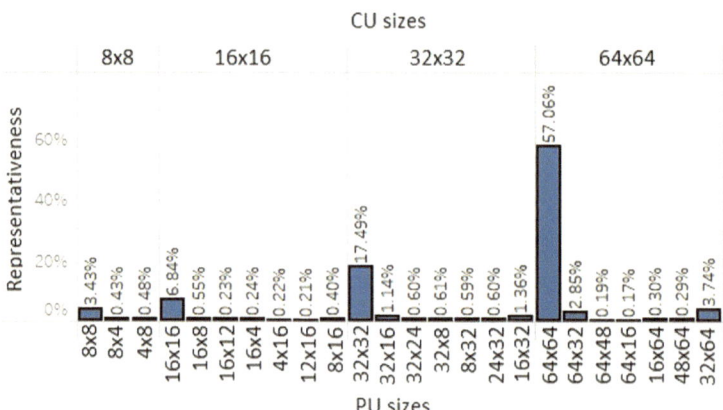

Fig. 3.37 Representativeness of PU sizes for the inter prediction considering depth maps and FCO approach

square-shaped sizes (64×64, 32×32, and 16×16 PU sizes) are the three most representative sizes, whereas 8×8, 16×16, and 32×32 PU sizes are the three most frequent sizes. Furthermore, 64×64 PU size is the fifth most frequent, whereas the 8×8 PU size is the fifth most representative size.

Considering the presented results, one can conclude that the square-shaped PU sizes have the two most selected sizes (16×16 and 8×8) and they have the two most representative sizes (64×64 and 32×32), considering both the texture and the depth maps. Considering depth maps, one can conclude that the square-shaped PU sizes are the three most selected sizes (32×32, 16×16, and 8×8) and they have the three most representative sizes (64×64, 32×32, and 16×16). Finally, the four square-shaped PU sizes are the most representative sizes in each one of the CU-depth levels for both texture pictures and depth maps.

3.3   General Discussion About the Experimental Results

Comparing the results obtained for both CCO and FCO approaches, it is possible to observe some differences between these two encoding order configurations, mainly in the results related to the depth-map encoding.

One of the main differences between CCO and FCO approaches is related to the usage of depth intra tools to encode the depth-map PUs, where only 6.61% of the PUs are encoded using depth intra tools in a CCO approach (considering both the texture pictures and the depth maps) and 23.31% of the PUs are encoded with depth intra tools in an FCO approach. This way, the percentage of pixels covered by depth intra tools increases from 3.62 to 8.95% when the CCO is changed to the FCO approach. Part of this variation is due to the increase in the number of pixels encoded using DIS mode in depth maps, which increases from 81.3% in the CCO approach to 86.68% in the FCO approach. The modification in the depth encoding process behavior is expected since the FCO dependent views encode depth maps before the texture pictures and, therefore, the encoder tends to select Intra-predicted candidates rather than using Inter-predicted candidates obtained with Merge mode or from disparity vectors.

Another important observation is related to the inexpressive use of N × N partitions in the depth-map coding in CCO and FCO where less than 1% of the depth-map pixels use this type of partition. However, the results also show that the N × N partition is more important in an FCO approach. Considering only the depth-map pixels and disregarding the DIS mode, the N × N partition increases the encoded pixels from 1.48% in a CCO approach to 6.06% in an FCO approach.

The distribution of pixels according to the CU-depth level in the depth-map encoding is also different in the FCO approach when compared with the CCO approach. Whereas CCO encodes more pixels in bigger CU sizes, the FCO approach encodes the depth maps in a more balanced way in terms of CU-depth levels, ranging from 14.85 to 38.37% according to the CU-depth level. In CCO configuration, this range is from 5.73 to 44.35%. Similarly, the usage of Intra modes in the depth-map encoding is different in an FCO approach. Whereas both DMM-1 and DMM-4 modes are largely used in a CCO approach, in an FCO approach, the DMM-1 mode tends to be more representative than the DMM-4 mode. In a CCO approach, the DMM-4 and the DMM-1 are the third and the fourth most selected modes among the Intra modes for depth maps, encoding 17.22 and 13.95% of the pixels, respectively. Considering an FCO approach, the DMM-1 is the second most selected mode, and the DMM-4 is only the fifth most selected mode among the Intra tools for depth maps, encoding 21.80 and 3.36% of the pixels, respectively. This different behavior of the DMM-4 encoding tool is also expected since DMM-4 requires the use of collocated texture samples which makes it impossible to use considering the dependent views in an FCO approach.

Focusing on the Inter prediction of texture pictures, the behavior in terms of block size selection and representativeness is practically the same in both FCO and CCO approaches, evidencing the importance of the square-shaped block sizes where the 8 × 8 and the 16 × 16 block sizes are the most often selected sizes and the 32 × 32 and the 64 × 64 block sizes are the most representative sizes for both FCO and CCO approaches disregarding the Skip mode. Focusing on the Inter prediction of depth maps, the behavior in terms of block size selection and representativeness is a little bit different for FCO and CCO approaches. The importance of the smaller block sizes is higher in the FCO approach, so the two most selected block sizes are the 8 × 8 and the 16 × 16, with 32.02 and 15.98%, respectively. The depth-map encoding with Inter prediction showed the importance of the square-shaped block sizes (64 × 64, 32 × 32, 16 × 16, and 8 × 8) in the coding process so that at least three square-shaped block sizes are the most selected and the most representative sizes for both coding order configurations disregarding the Skip mode. Notice that there are 24 PU sizes in the Inter prediction.

The ME importance is evident for both CCO and FCO approaches, where more than 80% of the Inter-predicted PUs in CCO and more than 90% of the Inter-predicted PUs in FCO use this type of prediction (disregarding Skip mode). DE is also important, being part of at least 20.62% of the pixels encoded considering the depth maps in the CCO approach and at least 10.7% of the pixels encoded considering the depth maps in the FCO approach. Notice that there is a significant decrease in the use of the DE considering the FCO. Such behavior can also be observed in the texture encoding, and it is related to the coding order of the dependent views, which tend to choose intra modes rather than using Merge mode and DE. As previously explained, some Merge candidates are provided by techniques that depend on the channel coding order. This way, the configuration is directly related to the choice of the modes. Note that DE stills to be important to the encoding of regions with occlusion and disocclusion.

Focusing on the texture behavior, the use of DBBP mode in an FCO approach also varies a lot. Whereas in a CCO approach, DBBP encodes 4.34% of the texture pixels considering only the Inter prediction and disregarding the Skip mode, in an FCO approach, DBBP is practically not used, encoding only 0.13% of the pixels considering the same scenario. Considering the fractional ME and DE, the significance of the quarter-pixel vectors is evident in the texture encoding process since 56.54% of the pixels are encoded using quarter-pixel FME and 16.54% of the pixels are encoded with quarter-pixel DE considering the CCO approach and disregarding the Skip mode. Considering the FCO approach, the importance of the quarter-pixel prediction is also notable. In the texture encoding process, 63.45% of the pixels are encoded using quarter-pixel ME, and 8.89% of the pixels are encoded with quarter-pixel DE, disregarding the Skip mode. It is important to highlight that depth maps do not use fractional-precision vectors.

Intra-Frame Prediction Developed Architectures

4

This chapter presents in detail the four high-throughput architectures developed for the Intra-frame prediction. The first architecture consists of a Depth Intra Skip (DIS) hardware design that adopts a distortion metric replacing, and data reuse in order to obtain energy efficiency. The second hardware design consists of a system capable of dealing with both the novel depth-map Intra-frame prediction modes, and the modes inherited by the HEVC standard, while reducing energy and memory requirements. The third architecture consists of an implementation of the novel 6WR algorithm. The fourth arquiteture consists of a flexible coding order implementation that exploits the correlations between angular modes and DMMs. All architectures used pipeline and the clock-gating low-power technique in order to reach processing rates capable of processing high and ultra-high definition videos and, also, to save energy consumption when it is possible. It is important to mention that the complexity-reduction heuristics proposed for the Intra-frame prediction in this chapter are anchored in the statistical results presented in Chap. 3. The impacts of the compression are considered in the CCO approach for the first three architectures and the FCO approach for the last architecture. However, all the algorithms and architectures can be used along an FCO approach without drastic modifications.

4.1 Low-Power Depth Intra Skip Architecture based on Distortion Metric Replacing

The architecture presented in the sequence was the first dedicated hardware design for the 3D-HEVC DIS coding tool published on the literature and it was developed considering all possible block sizes. A complexity-reduction strategy to obtain a low-power and

V. Afonso et al., *Hardware Design for 3D Video Coding*, Synthesis Lectures on Engineering, Science, and Technology, https://doi.org/10.1007/978-3-031-80232-4_4

high-throughput architecture was implemented to process the similarity criterion. Also, strategies of data reuse were used in the hardware design. Such strategies were adopted based on evaluations performed with the 3D-HTM reference software, as follows.

4.1.1 3D-HEVC DIS Complexity-Reduction Strategy and Evaluation

As presented in Sect. 2.1.5.11, SVDC requires a considerable number of arithmetic operations to decide the best mode to encode a given block. Furthermore, SVDC also imposes multiplications and costly rendering processes for hardware implementations. This way, replacing the SVDC by other similarity criterion that allows a hardware-friendly design is a promising complexity-reduction strategy. The use of SAD (see Sect. 2.1.1.3) as the similarity criterion directly applied to the depth-map samples avoids both multiplications and the rendering process.

Comparing (2.3) and (2.19) mathematically represent the SAD and the SVDC, respectively, it is possible to notice that the total arithmetic operations can be reduced at least 71.52% (depending on the block size) using SAD as similarity criterion. For instance, considering a 64×64 block, the total arithmetic operations can be reduced of 28,544 to 8128. Table 4.1 presents in details the total arithmetic operations as a function of the block size for the SAD and SVDC distortion metrics, given an $N \times N$ block, where N represents the block height/width.

Despite replacing SAD as the similarity criterion rather than using SVDC being a promising complexity-reduction strategy, the trade-off between coding efficiency and computational effort must be carefully considered. This way, the coding-efficiency impact of this strategy was evaluated with the 3D-HTM reference software in the version 16.0 (3D-HEVC Reference Software 2024), the IO temporal configuration and the test conditions recommended in the CTC document by the JCT-3V and presented in Appendix A. Therefore, the eight 3D video sequences were used, including five 1920×1088 sequences and three 1024×768 sequences as well as the four QP sets for texture/depth defined in the CTC document. Furthermore, it is worth saying that between 200 and 300 AUs were encoded according to these video sequences.

Table 4.1 Total arithmetic operations for an $N \times N$ block according to the distortion metric

Arithmetic operations	distortion metric		Δ (SVDC – SAD)
	SAD	SVDC	
Subtractions	N^2	$3N^2$	$2N^2$
Sums	$N^2 - N$	$2N^2 - 2N$	$N^2 - N$
Multiplications	–	$2N^2$	$2N^2$
Total	$2N^2 - N$	$7N^2 - 2N$	$5N^2 - N$

Table 4.2 presents the coding-efficiency degradation results for this analysis regarding BD-rate. Throughout this book, the coding efficiency results are mainly discussed for video total and synthesized frames. When *Video Total* is referred, the quality related to texture channel (disregarding synthesized frames) considering the total bit rate of the video (texture + depth) is measured. When *Synthesized Frames* are referred, the quality of frames synthesized with encoded texture and encoded depth maps, using DIBR, is measured in comparison with frames synthesized with original texture and original depth maps. Occasionally, quality results are also presented for *Video Only*, which represents the quality to the texture channel using the texture bit rate of the video (disregarding synthesized frames). According to the results presented in Table 4.2, using the SAD as the similarity criterion brings negligible losses at the 3D-HEVC coding efficiency, increasing the BD-rate only by 0.21% for synthesized views, on average. When the video total is considered, the result is similar, with a 0.19% BD-Rate increase. It should be noticed that these compression losses are acceptable when a significant computational effort is saved.

Note that the complexity-reduction strategy presented in this section is capable of decreasing the computational effort of the DIS coding tool while providing a negligible coding loss in terms of compression efficiency. Since the SAD similarity criterion is friendlier than the SVDC to hardware design, a dedicated 3D-HEVC DIS architecture, which supports the four CU sizes, is presented in the next subsection.

Table 4.2 BD-rate increase using SAD as similarity criterion

Test sequences	BD-Rate increase using SAD rather than SVDC (%)	
	Video total	Synthesized only
Balloons	0.16	0.23
Kendo	0.22	0.34
Newspaper_CC	0.22	0.26
GT_Fly	0.14	0.20
Poznan_Hall2	0.40	0.29
Poznan_Street	0.10	0.14
Undo_Dancer	0.07	0.02
Shark	0.17	0.21
1024×768	0.20	0.28
1920×1088	0.18	0.17
Average	0.19	0.21

4.1.2   Developed 3D-HEVC DIS Architecture

The developed architecture was designed to process the four block sizes supported by the DIS coding tool in parallel. For that, four DIS modules process and deliver the best SADs and DIS modes, one for each CU size, as presented in Fig. 4.1. In order to reach a high parallelism level, the proposed design was made in a bottom-up approach, where the DIS modules that process CUs bigger than 8 × 8 pixels compose their CUs by the processing of 8 × 8 blocks.

Note that the bottom-up processing strategy allows a similar hardware design for all DIS modules. The main difference among these modules is in the internal control module and the internal architecture bit depth. Once this strategy was adopted, it also provides a reduction of external-memory communication and computing by reusing input samples and partial SAD calculations. Therefore, this bottom-up processing strategy results in an efficient hardware design by allowing better control of the trade-off among hardware cost, power dissipation, and throughput.

Figure 4.2 shows the processing order (from 1 to 64) of 8 × 8 blocks aiming at composing bigger CU sizes. Note that this processing order follows a Z format, as highlighted in red (blocks 1, 2, 3, and 4), where a 16 × 16 block is composed by processing 8 × 8 blocks. Furthermore, the strategy of using a Z-format is extended to other CU sizes, i.e., 32 × 32 blocks can be composed by 16 × 16 blocks (green), and 64 × 64 by 32 × 32 blocks (blue). The Z-format scheme was adopted to make the DIS hardware design compliant with other Intra-frame coding tools, since these modes have data dependence due to the use of left neighboring column and above neighboring row as reference samples. It is worth saying the adopted approach makes feasible the global decision on which Intra tool presents better result based on RDO.

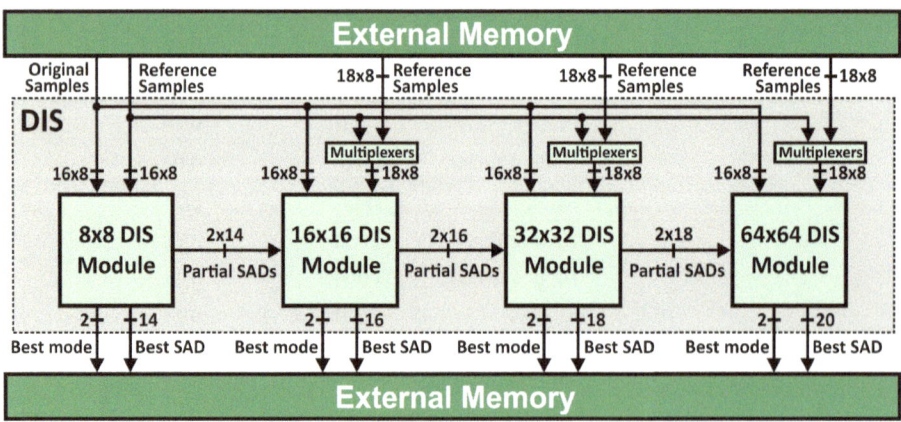

Fig. 4.1 Complete diagram for the developed DIS hardware design (Adapted from Afonso et al. (2017))

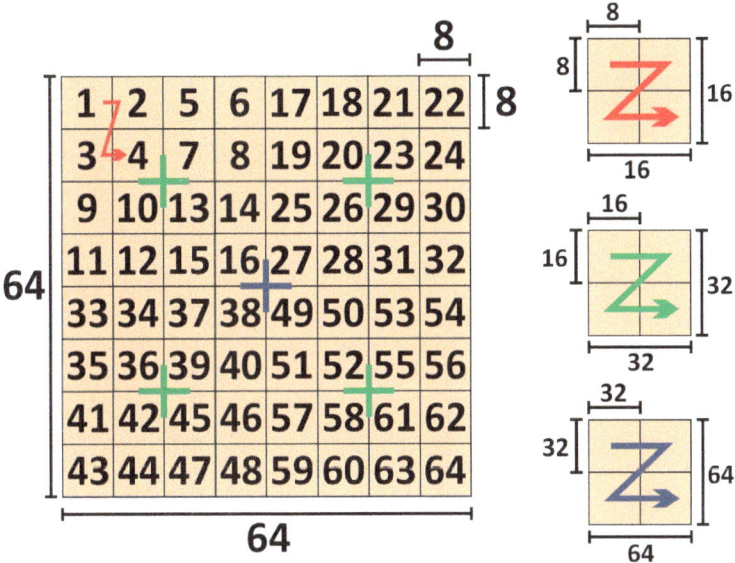

Fig. 4.2 Processing order of 8 × 8 blocks to compose the bigger CUs (Adapted from Afonso et al. (2017))

The data-reuse scheme of bigger DIS block sizes by processing 8 × 8 blocks is detailed in Fig. 4.3. Reuse of SAD values is possible for the IP_H and IP_V modes, while reuse of input samples is possible for the IP_H mode. In Fig. 4.3, yellow squares represent the reuse of SAD values, brown squares represent the reuse of input samples, and white squares show the blocks where no data reuse is possible. Note that the reuse of SAD values reduces about 50% the arithmetic operations needed to process 16 × 16, 32 × 32, and 64 × 64 DIS block sizes when the IP_H and IP_V modes are considered. Also, the reuse of input samples reduces 28.91% the memory accesses considering the IP_H mode when applied to the 16 × 16, 32 × 32, and 64 × 64 DIS block sizes.

It is also important to notice that the Fig. 4.3 depicts the data-reuse scheme based on 8 × 8 blocks, but the reuse of SAD values can be provided by any DIS Module that processes smaller CUs. However, using the SAD values from the immediately smaller module, unnecessary computing is saved. This way, the reuse of input samples is based on 8 × 8 blocks while the reuse of SAD values is based on the immediately smaller module.

In Fig. 4.4, it is shown a generic block diagram that represents any DIS Module. In general, the differences among all four DIS Modules (one for each CU size) are in the control of each module, the I/O bit depth of partial SADs, the internal bit depth from the Accumulators, and the Comparator. For example, the 8 × 8 DIS Module does not apply data reuse; therefore, this module has no partial SAD inputs. Anyhow, it provides partial

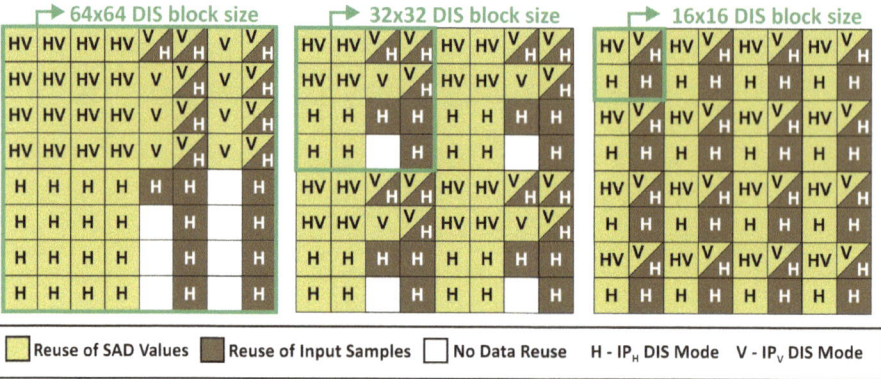

Fig. 4.3 Data-reuse scheme for 16×16 DIS blocks (left), 32×32 DIS blocks (center), and 64×64 DIS blocks (right; Adapted from Afonso et al. (2017))

SADs and reference samples for the DIS Modules that process bigger CUs. Additionally, each DIS Module has two Buffers, one to store horizontal reference samples and another to store vertical reference samples. Thus, these reference samples are selected by the DIS Modes stage using multiplexers.

DIS Modes stage delivers the outputs of the four prediction modes in parallel, row after row or column after column (both with eight samples of 8 bits) according to the mode. In the following, four SAD Trees compute – in parallel – the difference between

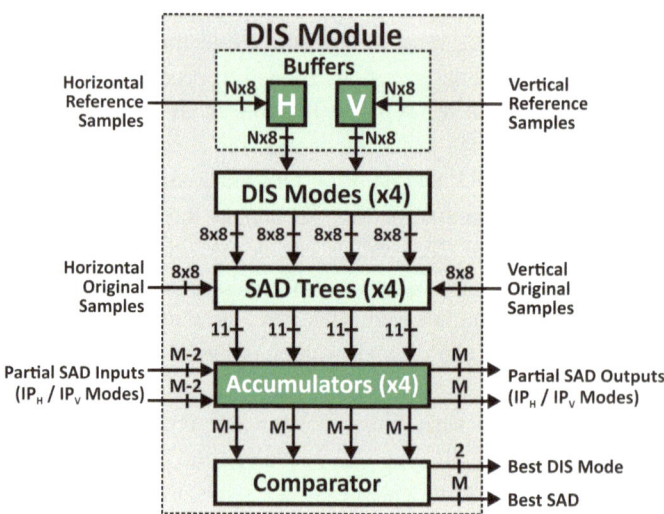

Fig. 4.4 Generic diagram of the DIS Modules (Adapted from Afonso et al. (2017))

the samples predicted by the four DIS Modes and the original samples for one row or column (according to the DIS prediction mode). Then, the Accumulators accumulate the obtained SAD of each row/column until computing the SAD of the entire block. Inside of the Accumulators stage, there are multiplexers which allow the reuse of SAD from smaller DIS Modules as well as partial SAD outputs that provide the SAD results for other greater DIS Modules. Given the data-reuse scheme, the clock-gating technique is applied to the DIS Modes, SAD Trees and Accumulators stages of the 16×16, 32×32, and 64×64 DIS Modules.

Finally, the Comparator compares the obtained SAD from the four DIS prediction modes, delivering the best SAD and the best mode for each processed CU. The complete encoding of an 8×8 block requires 13 clock cycles. In Fig. 4.4, **M** can assume values as 14, 16, 18, and 20 (bit depths), depending to the DIS Module whereas **N** can assume two values: eight or nine (samples) depending to the DIS Module as well.

4.1.3   Synthesis Results and Comparisons

The synthesis results were generated targeting an ASIC technology according to the experimental setup presented in Appendix B. These results are detailed in Table 4.3. It should be noticed that the hardware design presented in this book was the first dedicated solution for the 3D-HEVC DIS coding tool. Since direct comparisons with other works are not possible due to this fact, synthesis results of two different versions of the developed architecture were generated and compared. These two versions are called here as Non-optimized and Optimized. Both the Non-optimized and Optimized versions use the same hardware structure based on SAD as the similarity criterion presented in Sect. 4.1. The difference in terms of implementation is that the Non-optimized version does not implement the data-reuse scheme (SAD values and input samples) proposed in this book neither low-power techniques. This way, the Non-optimized version is used as a baseline to better evaluate the data-reuse and low-power strategies.

The maximum frequency obtained by the two different versions is superior to 550 MHz. This operation frequency is enough to process 60 UHD 2160p (3840 × 2160

Table 4.3 Synthesis results for ASIC technology

Architecture		Non-optimized	Optimized
Total area (gates)		34.80k	35.37k
1080p@120fps (five views)	Freq. (MHz)	252.72	252.72
	Power (mW)	14.95	11.52
2160p@60fps (five views)	Freq. (MHz)	505.44	505.44
	Power (mW)	24.64	19.57

pixels) fps considering five views. Since the two versions present the same parallelism, both versions reach real-time processing with the same operation frequency for HD and UHD resolutions. Another common point related to both versions of the architecture is the BD-Rate increase of about 0.21% in the synthesized views, on average. Note that this BD-Rate increase is acceptable since a high amount of computations was saved (at least 71.52%).

The Optimized version adopts the reuse of SAD values and input samples as presented in Fig. 4.3. Also, it implements the clock-gating register technique. As Table 4.3 shows, the Optimized version reduces the power dissipation in 20.58% considering 2160p@60fps resolution (five views) when compared to the Non-optimized version, reaching a power dissipation of 19.57 mW. Considering the processing of 1080p@120fps resolution (five views), the power-dissipation reduction achieves 23.05%. The hardware resource usage of the two versions is similar. However, it is important to highlight that the data-reuse scheme and the clock-gating technique increase by 1.09% the hardware resource usage. It is important to notice that this power-dissipation reduction is due to the data-reuse scheme and low-power techniques. The power-dissipation reduction due to the use of SAD as the similarity criterion rather than using SVDC was not measured in this book. By using SAD as the similarity criterion a costly rendering process related to the SVDC is also avoid.

4.2   Low-Power and Memory-Aware Depth-Map Intra-Frame Prediction System Based on Complexity Reduction Heuristics

The development of hardware designs for the novel Intra-frame prediction modes have some gaps that remain uncovered among the works published in the literature, as follows: (i) there are no works that implement both the novel 3D-HEVC Intra-frame prediction modes and the modes inherited from HEVC; (ii) some works are not fully compliant with the 3D-HEVC standard; (iii) DMM-1 hardware designs usually store all possible wedgelets with a significant cost in the memory-related energy (Sanchez et al. 2016). In this book, an intensive effort is done in order to overcome these gaps through the development of a low-power and memory-aware depth-map Intra-frame prediction system, as follows. The next subsection presents complexity-reduction strategies and their evaluation and, after, another subsection presents the hardware design.

4.2.1   3D-HEVC Depth-Map Intra Prediction Heuristics and Evaluation

Section 3.2.1 presented a comprehensive statistical analysis of the 3D-HEVC depth-map intra-prediction tools usage. Among the data provided by the experiments, the most prominent modes and block sizes used in the depth-map coding were identified.

The statistical analysis of the intra-prediction tools usage was based on the percentage of pixels that are encoded using each encoding tool, mode and block size. Considering the CCO approach, the 3D-HTM depth-map coding used the DIS tool to encode 81.13% of the depth-map pixels, i.e., all the remaining 37 prediction modes are responsible for 18.87% of the pixels. Then, the high importance of the DIS tool becomes evident. But this fact cannot lead us to consider that the other encoding tools are less important than the DIS, since the other tools better deal with sharp edges and directional structures which are extremely important to the view synthesis process.

Since the DIS tool is much more used than the others, the Intra-frame prediction distribution was also analyzed disregarding the DIS tool in Sect. 3.2.1. Considering the remaining 37 prediction modes, the Planar mode is the one with higher representativeness, encoding 29.33% of the depth-map pixels, followed by the DMM-4 mode with 17.22%, the DMM-1 mode with 13.95%, and the DC mode with 6.53%. The fifth mode with higher representativeness is the Vertical mode, encoding 3.81% of the depth-map pixels, followed by the Horizontal mode with 1.81%. If these six modes are considered together, they are responsible for encoding 72.65% of the depth-map pixels when DIS is not considered.

Regarding the block sizes used for the depth maps in the Intra-frame prediction, the 4×4 block size is unrepresentative, corresponding to only 0.28% of the pixels in the depth-map coding. The results show that the bigger block sizes encode a higher amount of depth-map pixels when compared to the smaller sizes. Disregarding the DIS tool, the 4×4 block size encodes only 1.48% in this case whereas 32×32 and 64×64 block sizes have representativeness in the encoding of 36.09 and 44.35% of the depth-map pixels, respectively. The 16×16 block size encodes other 13.83% whereas the 8×8 block size encodes 5.73% of the depth-map pixels disregarding the DIS tool.

Based on the analyses presented in Sect. 3.2.1, some hardware-oriented heuristics are proposed in this book in order to reduce the computational effort for the depth-maps Intra-frame prediction. These strategies consist in to remove the less important prediction modes and block sizes and in a specific strategy applied to the DMM-1 mode.

The depth-map Intra-frame prediction was restricted to the seven most representative prediction modes and the four most representative block sizes considering the CCO approach, which corresponds to 94.84% of all encoded depth-map pixels considering the regular encoding process. Therefore, DIS, Planar, DC, Horizontal, Vertical, DMM-1, and DMM-4 modes and 8×8, 16×16, 32×32, and 64×64 block sizes were selected to be supported in the hardware design.

Furthermore, other simplifications were introduced to the DMM-1 mode in order to reduce its processing and memory requirements, since this is the most computational intensive mode among the supported ones. The used heuristic reduces the number of wedgelets tested during the encoding process to six. The six wedgelets are defined using a pre-processing through a gradient calculation using the samples along the four neighboring-block borders. Then, all DMM-1 wedgelets are available, but only six will be evaluated. The heuristic also avoids memory usage for wedgelet patterns storage using the Bresenham (Bresenham 1965) algorithm to compute at run-time the DMM-1 bitmaps for the six evaluated wedgelets. The Bresenham is widely used in Computer Graphics to perform the rasterization of lines and polygons. A modified Bresenham implementation receives the start and end points of the wedgelet and processes a pixel of the line representing the wedgelet at the time. Then, the 1908 bitmap wedgelet patterns that must be stored when using the conventional approach were reduced to only six evaluations without any storage.

In order to verify the impact of the developed heuristics on the coding efficiency, these hardware oriented heuristics were inserted in the 3D-HTM in version 15.1 (3D-HEVC Reference Software 2024) and compared with the original results using the same test conditions presented in Appendix A. Therefore, the experiments considered the 3D-HEVC CTC document, i.e., the eight videos at 1920×1088 and 1024×768 resolutions and four QP sets. For this evaluation, two parameters are considered, as follows: (i) the encoding time, used to estimate the computational effort reduction; and (ii) the BD-rate parameter, to evaluate the bit-rate variation for the same image quality.

Considering the average of all videos and QPs running under RA temporal configuration, the BD-Rate increased 2.64% in the synthetic views, on average, as presented in Table 4.4. On the other hand, the computational effort was reduced in 32.82% when encoding depth maps and in 16.17% for the global 3D-HEVC encoder. Regarding the IO temporal configuration, i.e., only intra-prediction tools are used in the encoding process, the simplifications increase the BD-Rate by 7.16% in the synthetic views, on average. In this case, the encoding-time of depth maps is reduced by 53.98% and the total encoding time is reduced by 46.58%, as shown in Table 4.4.

4.2.2 Developed 3D-HEVC Depth-Map Intra Prediction Architecture

This subsection presents the architecture designed for the depth-map Intra-frame prediction using the heuristics presented in the last subsection. The architecture was designed to efficiently process the modes Planar, DC, Horizontal, Vertical, DIS, DMM-1, and DMM-4, supporting 8×8, 16×16, 32×32, and 64×64 block sizes. The heuristic to simplify the DMM-1 mode was also considered.

Table 4.4 BD-rate variation by using the hardware-oriented heuristics

Sequence	IO temporal configuration				RA temporal configuration			
	BD-rate variation		Encoding-time reduction		BD-rate variation		Encoding-time reduction	
	Video total (%)	Synthesized (%)	Total (%)	Depth only (%)	Video total (%)	Synthesized (%)	Total (%)	Depth only (%)
Balloons	−0.02	7.96	45.86	53.26	0.22	2.73	14.95	32.21
Kendo	0.02	8.04	46.29	53.81	0.61	2.31	14.88	31.13
Newspaper_CC	−0.89	12.09	48.35	54.46	−0.32	6.48	18.01	35.03
GT_Fly	−0.61	3.82	46.68	54.01	−0.05	0.88	15.48	31.92
Poznan_Hall2	−0.29	7.98	46.39	54.69	−0.15	2.32	17.01	33.94
Poznan_Street	−0.39	3.31	47.27	53.88	0.00	2.13	18.57	35.61
Undo_Dancer	−0.18	7.55	45.20	53.45	0.53	2.28	14.56	31.35
Shark	−0.87	6.58	46.64	54.30	0.23	1.98	15.89	31.34
1024 × 768	−0.30	9.36	46.83	53.84	0.17	3.84	15.94	32.79
1920 × 1088	−0.47	5.85	46.44	54.07	0.11	1.92	16.30	32.83
Average	−0.41	7.16	46.58	53.98	0.14	2.64	16.17	32.82

The designed architecture is presented in Fig. 4.5 and it is divided into Processing Cores (PC), where each PC is responsible for processing a subset of the pre-defined intra-frame prediction modes in an interleaved way, as presented in Fig. 4.5. Then it is possible to reach the desired throughput, avoiding unnecessary calculations.

The seven supported modes are distributed in three processing units that work in parallel, as presented in Fig. 4.5. The intra tools are grouped according to their similarities, which allows a reduction in area usage (and power dissipation) by reusing the hardware inside each PC. In the high-level diagram of the architecture presented in Fig. 4.5, one can see the mode distribution on the three PCs: (i) Planar, DC and DMM-4; (ii) IP_H, IP_V, SD_H, SD_V; (iii) DMM-1. The IP_H and IP_V modes are used both as HEVC conventional modes as DIS sub-modes, allowing the reuse of PC-2 results. Although the DMM-1 and DMM-4 modes have similarities, these modes were not grouped in the same PC in order to obtain the desired throughput, since DMM-1 is much more complex than the other modes.

The designed hardware is a multiplierless solution, where the multiplications were replaced by shift-adds to save hardware resources (and power). Clock-gating was also widely used to idle modules during the architecture operation. The architecture has a parallelism level enough to deliver one block line per clock cycle independently of the block size.

The Processing Core-1 (PC-1) is responsible for processing the Planar, DC, and DMM-4 modes. Each mode is also processed by an independent module, as one can observe in Fig. 4.5. As inputs, the PC-1 receives the left and upper edges of neighboring blocks to the block being encoded (previously reconstructed blocks), which are used for DC and Planar calculations. For the DMM-4 calculations, the PC-1 architecture also receives four 8-bit width samples from the texture, which are required for the threshold calculation. PC-1 also receives as inputs the samples of the depth-block being predicted to calculate the residues. As one can see in the Timing Analysis presented in Fig. 4.6, the residue calculation starts at the moment that the Planar mode evaluates the first line of samples. The DC calculator needs one cycle before the residue calculation.

The PC-2 is responsible for calculating the IP_H, IP_V, SD_H and SD_V modes and each mode is processed by an independent module, as presented in Fig. 4.5. As inputs, PC-2 receives the left and upper edges of neighboring blocks to the block being encoded (previously reconstructed blocks). Also, the PC-2 receives as inputs the samples of the block being predicted to calculate the residues. A multiplexer selects which samples can follow to the Residue Calculator at each moment. The coding of the four modes starts along with the residue calculation, avoiding any additional clock cycle besides the ones used by the Residue Calculator, as presented in Fig. 4.6.

The PC-3 is responsible for calculating the DMM-1. As discussed in the last subsection, a heuristic to drastically reduce the DMM-1 computational effort and to avoid the storage of wedgelet patterns in memory was used. The heuristic consists of processing only the six wedgelets formed from the four higher gradients along the four borders of the block to be predicted (one gradient per border), as detailed in the example given in

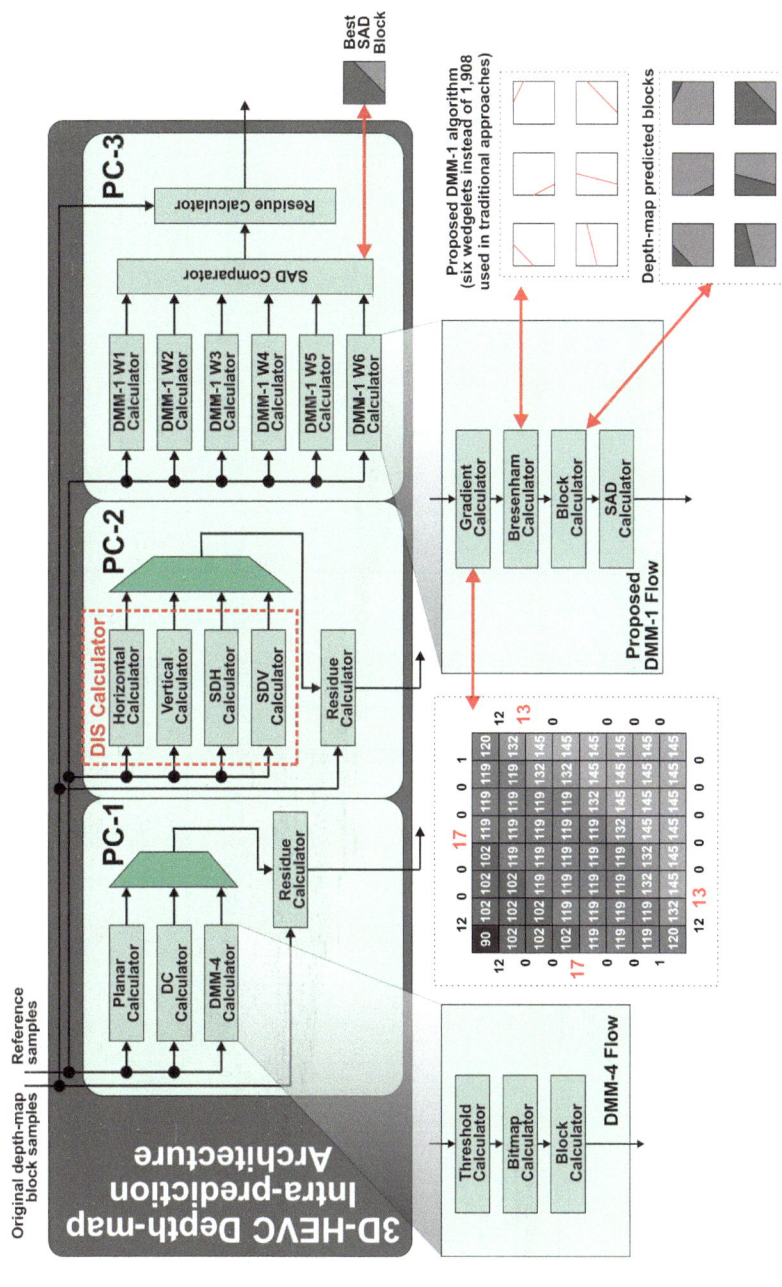

Fig. 4.5 High-level block diagram of the developed architecture (Adapted from Ücker et al. (2020))

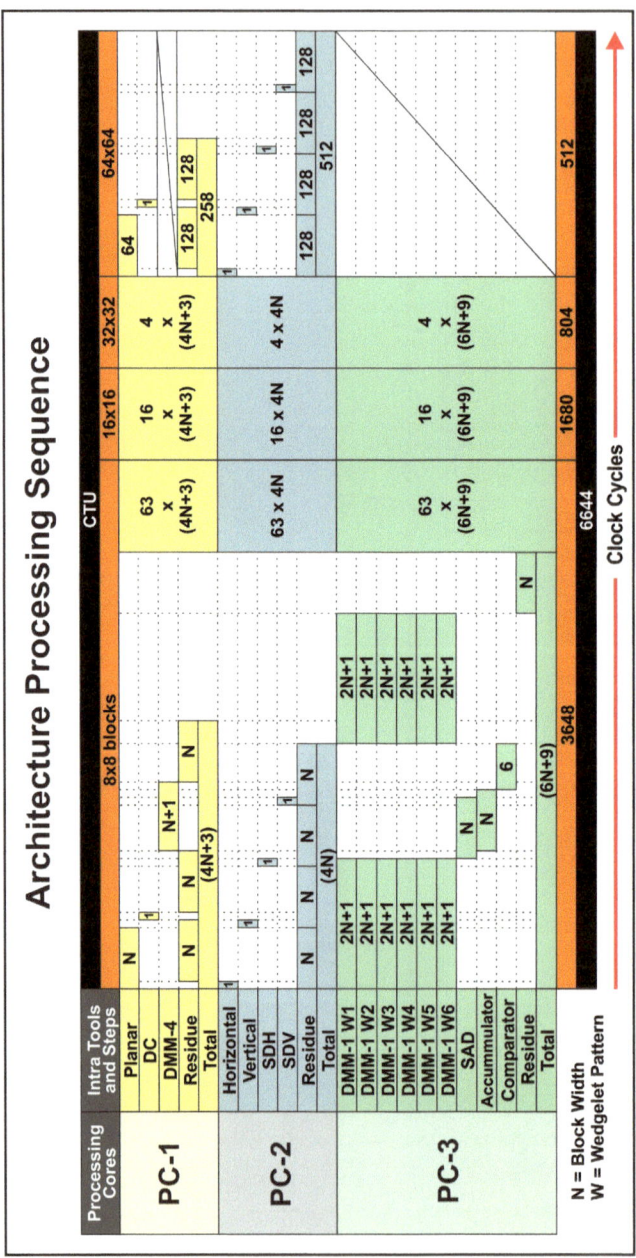

Fig. 4.6 Timing analysis of the developed architecture (Adapted from Ücker et al. (2020))

Fig. 4.5. PC-3 processes these six wedgelets in parallel, as presented in Figs. 4.5 and 4.6. As inputs, PC-3 receives the samples of the block being predicted to calculate the residues.

The first DMM-1 module consists of a Gradient Calculator that generates six outputs determining the start and end points of the six wedgelets. Once the positions to compose the wedgelet are available, the bitmap of the predicted block can be obtained using the Bresenham algorithm in order to indicate which samples belong to each partition. It is important to notice that the Bresenham algorithm does not predict the number of samples the line will have. Thus, it is not possible to predict the total number of required clock cycles. The architecture was designed to cover the worst case of the Bresenham algorithm. The next step is the Block Calculator that starts its operation processing the bitmap line by line. After the Block Calculator generates the predicted block lines, the SAD calculator generates the sum of absolute differences related with each one of the six evaluated wedgelets. Finally, PC-3 has a comparator to compare the six SAD values and decide the most efficient wedgelet before calculating its residue.

4.2.3   Synthesis Results and Comparisons

The developed architecture was synthesized targeting an ASIC technology according to the experimental setup presented in Appendix B. Table 4.5 summarizes the hardware results in the rightmost column. Table 4.5 also presents comparisons with related works. This hardware design was the first to process depth maps that supports both the novel 3D-HEVC Intra-frame prediction modes and the HEVC inherited Intra-frame prediction modes. Although the literature presents some works with hardware designs targeting the HEVC Intra-frame prediction modes, those works do not consider the MVD approach, making unfair the comparisons with the architecture presented in this book.

Among the related works targeting 3D-HEVC intra-prediction tools, three hardware designs found in the literature were selected for comparisons. It is important to emphasize that the work (Sanchez et al. 2014b) implements only the DMM-4, the work (Sanchez et al. 2016) implements both DMM-1 and DMM-4, and the work (Amish et al. 2019) implements DIS, DMM-1, and DMM-4. It is also important to emphasize that all the works (Sanchez et al. 2014b, 2016; Amish et al. 2019) are not fully compliant with the 3D-HEVC standard.

As one can observe in Table 4.5, only the architecture presented in this book processes HEVC inherited modes (Planar, DC, Horizontal, Vertical) besides the DIS, DMM-1 and DMM-4 modes. Whereas the works (Sanchez et al. 2014b; Amish et al. 2019) focused on FPGA devices, this architecture and the work (Sanchez et al. 2016) are focused on ASIC technology. This hardware design and the other ASIC-based work present power results and even the hardware developed in this book supporting a high number of prediction modes, it presents promising results.

Table 4.5 Synthesis results for ASIC technology

Work	Sanchez (2014b)	Sanchez (2016)	Amish (2019)	Developed architecture
Encoding tools	DMM-4	DMM-1 and DMM-4	DIS, DMM-1, and DMM-4	DIS, DMM-1, DMM-4, H, V, Planar, and DC
Technology	FPGA Altera StratixV 5SGXMABN3F45I32 (28 nm)	ASIC 65 nm	FPGA Xilinx Virtex 6 (40 nm)	ASIC Nangate 45 nm
Area	5568 ALMs 4405 Registers	219.95 k gates* (2-input NAND)	55 k LUTs 67 k Registers	486,804 gates (2-input NAND)
Memory for DMM-1 wedgelets	–	Yes (101,352 bits*)	Yes (1947 k bits)	No
Maximum frequency	31.3 MHz	53.2 MHz*	275 MHz	940 MHz
Frequency for HD1080p@30@5views	22.47 MHz	Not reached	165 MHz	504.5 MHz
Power for HD1080p@30fps	Not available	166.5 mW* (1 view)	Not available	41.57 mW/0.95 V (9 views)
3D-video processing at UHD2160p@30fps	No (1.74 views)	No (0.25 views)	Yes (2.08 views)	Yes (2.33 views)

*Values consider only the architecture to process 32 × 32 blocks

As one can notice in Table 4.5, only this hardware design and the design presented in (Amish et al. 2019) are able to process 3D-videos at UHD2160p@30fps (with at least two views). However, the work (Amish et al. 2019) requires an external memory to store all the possible DMM-1 wedgelets patters, which is not necessary for the architecture presented in this book. Similarly, the work (Sanchez et al. 2016) also needs external memory. These works also did not present the power dissipation related to the memory used to store the wedgelets. The developed architecture can process 3D videos at HD1080p@30fps (nine views) with a power dissipation of 41.57 mW whereas the architecture developed by (Sanchez et al. 2016) needs 166.5 mW only to process the 32×32 blocks and one view at HD1080p@30fps. It is important to notice that the power dissipation of the architecture presented in this book covers the processing of all supported block sizes (from 4×4 to 64×64) whereas (Sanchez et al. 2016) does not present the total power dissipation by processing all block sizes.

Since the other works only support a few Intra-frame prediction modes and do not present the BD-rate results for the complete Intra-frame prediction, as this book presented, it was not possible to make a fair comparison of the BD-rate results of these works with the architecture developed in this book. Considering that the developed architecture covers both the novel Intra-frame prediction modes and the ones inherited from the HEVC, the BD-Rate increases 2.64% in the synthetic views under RA temporal configuration. The work (Sanchez et al. 2016) increases by 0.09% the BD-Rate in the synthetic views but it supports only DMM-1 and DMM-4 modes. The work (Amish et al. 2019) increases from 0.436 to 1.906% the BD-Rate in the synthetic views but it covers only DMM-1, DMM-4, and an older DIS implementation.

4.3 DMM-1 Energy-Aware and High-Throughput Hardware Design Based on 6WR Algorithm

Although the new intra-prediction tools introduced by the 3D-HEVC can be used to encode the depth maps efficiently, the computation of these tools causes a significant impact on the encoding time of the encoder. In fact, considering the All-Intra (AI) profile, where only the intra-prediction tools were available, the HEVC inherited modes demand 12.1% of the total encoding time, the DMM-4 and DIS modes together demand less than 9%, while the DMM-1 is the most complex intra-prediction tool from 3D-HEVC, since it demands at 27.2% of the total encoding time (Afonso et al. 2019b).

Several works can be found in the literature proposing dedicated hardware for the HEVC inherited tools, some of which can be used in a 3D-HEVC context with minimal changes, as (Tseng et al. 2019; Xu et al. 2018; Correa et al. 2017; Fang et al. 2015; Ramos et al. 2018). Also, there are some works focused on dedicated hardware for the new tools of the 3D-HEVC, as (Afonso et al. 2019a, 2017; Ücker et al. 2018; Sanchez et al. 2014, 2019). Although, there are only a few works completely focused on dedicated hardware to implement the most complex tool of depth map intra-prediction from 3D-HEVC, such as (Sanchez et al. 2014, 2019). The work presented by Sanchez et al. (2019) proposes a scalable encoder and decoder hardware for the DMM-1 and DMM-4 that can be used to evaluate any PU size supported by 3D-HEVC. The DMM-1 algorithm was slightly modified, where the refinement stage was removed to reduce its hardware resource and memory usage, resulting in a coding-efficiency loss of 0.09% regarding the Bjontegaard Delta Rate (BD-Rate; Bjontegaard 2008). The developed hardware has a power dissipation of 151.6 mW. However, it does not reach the necessary throughput to encode more than one view of HD1080p videos at 30 frames per second (fps). Besides, it is important to emphasize that the MVD format adopted in the 3D-HEVC requires the processing of more than one view to take advantage of all features (Merkle et al. 2007), as the DIBR process.

A dedicated hardware design for three Intra-Prediction modes (DIS, DMM-4, and DMM-1) was proposed in (Amish et al. 2019). The implementation is based on an algorithm that evaluates only one wedgelet to each PU size according to the gradients of the input block edges. This strategy led to a coding-efficiency loss of 1.115% in terms of BD-rate. However, the evaluations in (Amish et al. 2019) were not completely based on the Common Test Conditions (CTCs) document since they used only four of the eight sequences recommended by the CTC. Besides, the Random-Access (RA) profile was used, where the coding efficiency impact of intra-prediction modes can be masked by the inter-prediction. The developed hardware was synthesized only for FPGA technology, and despite it reaching the necessary throughput to process six views of HD1080p videos at 30 fps, no results of power dissipation were provided. Therefore, it is necessary to design hardware for the 3D-HEVC intra modes that are able to provide the necessary throughput to enable 3D video applications jointly with low power strategies to save battery in mobile devices.

In this book, we develop a dedicated high-throughput hardware design that implements a fast and energy-aware algorithm for the DMM-1 tool, exploring the gradients of the block edges. The simulation results in 3D-HEVC Test Model (3D-HTM; 3D-HEVC Reference Software 2024) show that the coding efficiency degradation is 2.819%, considering the AI profile. The ASIC synthesis results demonstrated that it is capable of processing up to nine views of a 3D full HD 1080p video (MVD format) in real-time at 30 fps, with a power dissipation of only 263.7 mW. This is the best energy-efficient architecture in comparison with the related works.

The DMM-1 intra-prediction mode was developed to encode the depth-map regions that contain sharp edges efficiently. The DMM-1 operates with PU sizes of 4×4, 8×8, 16×16, and 32×32 samples. To encode a block, the DMM-1 applies a straight wedgelet to divide the block into two regions. So, all samples that belong to each region are represented by the same value – typically, the average from all values of the original block that belongs to that region. Several wedgelets could be chosen to encode the block, selecting the wedgelet that results in the smallest Sum of Absolute Differences

(SAD) value. The next subsection explains the wedgelets generation and its evaluation as performed by the 3D-HTM.

4.3.1 The Proposed 6WR DMM-1 Algorithm

In the 3D-HTM 16.2, a list of wedgelets is generated at the beginning of the encoding process, and the wedgelets are stored in internal memory. The generation of each wedgelet is based on allocating a start and end positions of each wedgelet and dividing the block into two regions, as presented in Fig. 4.7. This allocation is different according to the block size (Ikai et al. 2015). The wedgelets of 4×4 and 8×8 blocks consider the start and end positions at every edge sample, while the wedgelets of 16×16 and 32×32 consider these positions at every two and four edge samples, respectively. Also, in the wedgelets generation, the wedgelets from the list are categorized for the main stage or the refinement stage of evaluation (Ikai et al. 2015), as explained in the sequence. Besides, the wedgelets that satisfy some conditions are not included in the list, like the ones that generate a very similar partitioning from other wedgelet that are already on the list. This approach results in a total of 1908 wedgelets that can be used to divide each block, where there are 86 possibilities for 4×4 blocks, 802 for 8×8, and 510 for 16×16 and 32×32 blocks (Ikai et al. 2015).

All those wedgelets must be evaluated to choose the one with the best coding efficiency (minimize both distortion and bitrate). In the 3D-HTM, the evaluation of the list of wedgelets is performed into two stages, named the main stage and the refinement stage (3D-HEVC Reference Software 2024; Ikai et al. 2015). The main stage evaluates all wedgelets related to the current block size that were categorized for the main stage. This way, the main stage evaluates 58 wedgelets for 4×4 block sizes, 314 for 8×8 blocks,

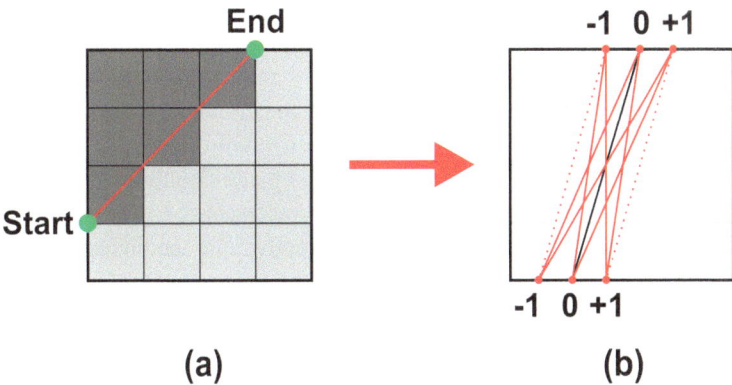

Fig. 4.7 DMM-1 prediction. **a** Wedgelets generation by allocating *start* and *end* positions; **b** wedgelets displacement in the refinement stage (Adapted from Perleberg et al. (2020b))

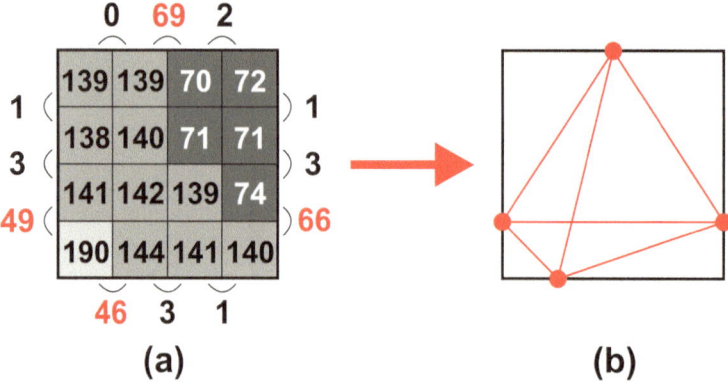

Fig. 4.8 Example of the proposed 6WR algorithm. **a** Gradient calculator in each edge; **b** six wedgelets to be evaluated in the main stage (Adapted from Perleberg et al. (2020b))

and 384 for 16×16 and 32×32 blocks. After the main stage, the refinement evaluates at eight wedgelets from the wedgelets classified for the refinement. These additional eight wedgelets from the refinement stage are similar to the result of the main stage, but they use the start and end positions displaced by ± 1 edge sample around the main stage best result (Ikai et al. 2015). Figure 4.7b presents the eight wedgelets to be refined (red lines) around the best result of the main stage (black line). Since some conditions restrict the generation of the refinement wedgelets, some wedgelets could generate the same partitioning. This way, the 3D-HTM implementation avoids unnecessary evaluations, which reduces the total of eight refinement wedgelets associated with its refinement stage. In fact, from the 1140 wedgelets available in the main stage, there is only an average of 1.8 refinement wedgelets associated with each wedgelet (3D-HEVC Reference Software 2024).

To reduce the DMM-1 complexity, we propose a fast and energy-aware DMM-1 algorithm, called 6WR (Six Wedgelets and six Refinements). The 6WR reduces the number of wedgelets evaluated in the main stage and also in the refinement stage to only six for each respective stage. The six wedgelets to be evaluated in the main stage were obtained based on the highest gradient of the edges of the current block. First, the algorithm obtains the highest gradient from the four edges of the block, by calculating the difference between the values of every two neighboring samples, as represented by the values around the block of Fig. 4.8a. All difference values are compared to find the highest difference value from each edge, represented in Fig. 4.8a by red values.

Then, the position of the highest difference from each edge of the block, represented by green circles in Fig. 4.8b, are combined to define the six wedgelets to be evaluated, represented in Fig. 4.8b by red lines. In the refinement stage, the reduction from eight to six wedgelets is performed by ignoring the two wedgelets that vary their start and end points to the same side (two dotted red lines in Fig. 4.7b), since these two wedgelets are the ones that result in the greatest variation among the two regions of the wedgelet. In conclusion, by the adoption of this algorithm, the number of evaluated wedgelets in the DMM-1 intra mode can be reduced from 94 (main stage + refinement stage) to 12 wedgelets for the 4 × 4 block size prediction, and a reduction from 810 (main stage + refinement stage) to 12 wedgelets for the 8 × 8 block size prediction.

4.3.2 Coding Efficiency Evaluation

The proposed 6WR algorithm was implemented in the 3D-HTM in version 16.2 (3D-HEVC Reference Software 2024) and evaluated according to the CTC (Müller et al. 2014) definitions (see Appendix A). The CTC document indicates eight video sequences to be used in evaluations, being each sequence composed of 200 to 300 frames, which corresponds to about 10 seconds of video. It also indicates three views of each video sequence, where each view is composed of both texture and depth map channels. Finally, four pairs (texture/ depth) of Quantization Parameters (QPs) were used: 25/34, 30/39, 35/42, and 40/45 (Müller et al. 2014).

The impact in the encoding efficiency of applying the proposed 6WR algorithm was investigated. This impact in the encoding efficiency was measured by using the BD-Rate metric (Bjontegaard 2008) to each sequence, considering the AI and the RA profiles. The BD-Rate represents the bit-rate variation of the proposed algorithm in comparison with the original one while keeping the same objective quality (PNSR), and it was obtained by considering the results achieved for each sequence with the four pairs of QP values.

The coding efficiency evaluation for all tested video sequences is presented in Table 4.6. The impacts considering the AI profile are more expressive than the ones reached with the RA profile. This is because in the AI only Intra-prediction tools are available, while the RA can use several Inter-prediction tools that could mask the impact of these modifications. Table 4.6 also presents the average BD-Rate value of each resolution. These average results show that the proposed 6WR algorithm presents a higher impact on the lower resolution sequences. This probably occur because each block has more image details in lower resolution, increasing the impact of the proposed algorithm. Finally, from Table 4.6 one can observe that the proposed 6WR algorithm presents an average increase in the BD-Rate of 2.819 and 1.162%, considering the AI and RA profiles, respectively.

The original DMM1 algorithm are responsible for the aver-age encoding time of 20.28%, while the 6WR algorithm is responsible for 0.48%, resulting in a encoding time reduction of 97.62% considering the complete CTC in AI profile. The runtime in RA profile was not measured since intra-prediction tools represent a small runtime proportion. Therefore, considering this encoding time reduction, the small impact in the encoding efficiency, and the reduction at 98.5% in the total number of evaluated wedgelets, the proposed algorithm is an excellent option to be adopted in an energy-aware dedicated hardware of the 3D-HEVC DMM-1.

Table 4.6 BD-rate increase

Resolution	Sequence	All-intra (%)	Random-access (%)
1024 × 768	Balloons	2.950	1.051
	Kendo	3.104	0.937
	Newspaper	4.883	2.614
	GT_Fly	1.537	0.453
	PoznanHall2	3.677	1.242
1920 × 1088	PoznanStreet	1.285	0.758
	UndoDancer	2.612	1.075
	Shark	2.502	1.164
Average 1024 × 768		3.646	1.534
Average 1920 × 1088		2.322	0.938
Average		2.819	1.162

4.3.3 The 6WR Hardware Design

The 6WR developed architecture is based on the solution presented in Fig. 4.9, where two unities like the one presented in Fig. 4.9, are instantiated, one for 4 × 4 PUs and other for the other PU sizes. The processing starts with the Gradient calculator, which computes the difference value between every two neighboring samples of all four edges of the input block, aiming to find the higher difference from each edge.

These four positions are grouped to generate the six wedgelets, as presented in Fig. 4.8. Then, each one of these wedgelets is processed by a different Wedgelet processor, which are presented in Fig. 4.9. Each Wedgelet processor generates the SAD value of a specific

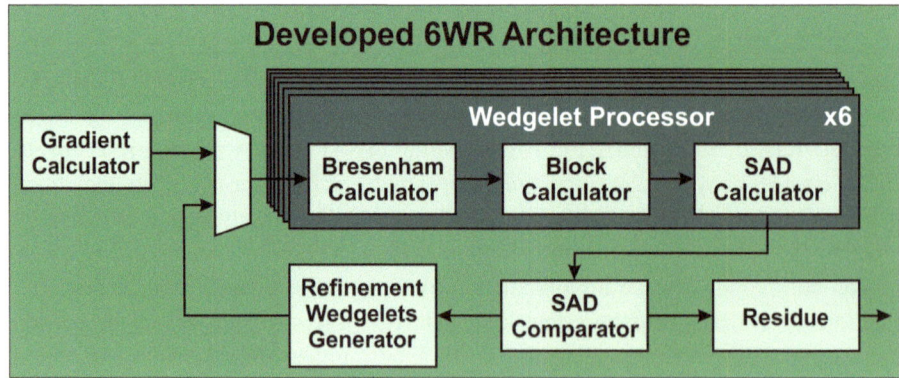

Fig. 4.9 Basic unit of the developed 6WR architecture (Adapted from Perleberg et al. (2020b))

predicted block. The first module of the Wedgelet processor is the Bresenham calculator, which generates the bitmap block used to divide the blocks into the two regions. The Bresenham algorithm (Bresenham 1965) is widely used in Computer Graphics to draw lines of objects, since it was an algorithm that can be applied without conventional multipliers, by only using shift-add operators. It was adopted in this architecture since it can compute the bitmap block at run-time, avoiding the use of internal memory to store the wedgelets. On the bitmap block, one region is represented by the value 0, while the other region is represented by the value 1. The bitmap block is used as input by the Block calculator.

The Block calculator is responsible for computing the predicted block. It was composed of two sets of accumulators used to obtain the sum of sample values and the number of samples that belong to each region. The value of the bitmap indicates to which region belongs each sample, so denoting in which set of accumulators the sample should be processed. Then, the Block calculator reconstructs the predicted block by using the average values of each region along with the bitmap block. For this processing, the value of the bitmap indicates that a sample from the predicted block must be composed of the average value of region one or two. After generating the predicted block, the SAD value of the predicted block is computed by the SAD calculator. This processing is performed line by line, so the SAD calculator accumulates the SAD value from all lines of the block together, aiming to obtain the SAD value of the predicted block. Finally, the outputs of the six Wedgelet processors are sent to the SAD comparator unit. It compares the SAD values of the six evaluated wedgelets, indicating the wedgelet with the smallest SAD value. Then, the Residue will generate the residue block of the best evaluated wedgelet.

In parallel to the Residue calculation, based on the information of the best wedgelet, six new wedgelets are evaluated to start the 6WR refinement stage. These six new wedgelets are obtained by varying the start and end position of the main stage results into ± 1 edge sample, as explained in Sect. 4.3.1. The information of these six new wedgelets is then processed by the Wedgelet processor, which will compute the SAD value of the six wedgelets. These six SAD values are then processed by the SAD comparator. Lastly, the residue of the best wedgelet is computed, ending the processing of the proposed algorithm for a given PU.

Figure 4.10 presents the timeline processing of the developed hardware. Since almost all modules from the architecture have several pipeline stages, performing its processing line by line, the delay from almost all modules are function of N, the size of the processed PU. From Fig. 4.10 one can notice that almost all modules were used two times to process each PU since they all are used to process both main and refinement stages of the algorithm. The residue was also computed two times for being used when the main stage result was best than the refinement wedgelets. Figure 4.10 also presents a timeline from the two units of the designed hardware that works in parallel, being one unit to deal only with 4×4 PU sizes, and one unit to deal with another PU sizes. Since there are

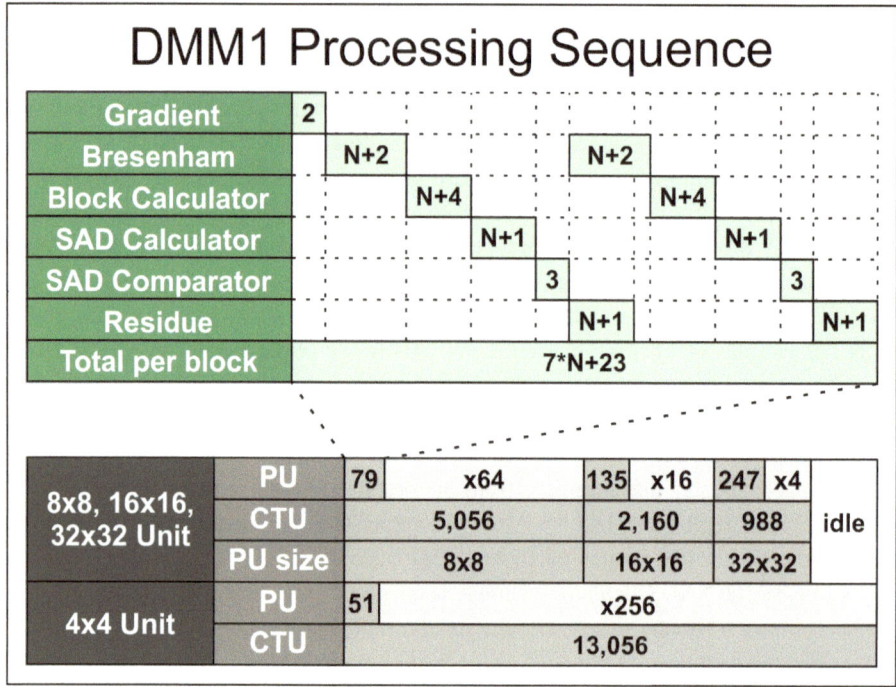

Fig. 4.10 The processing sequence from the DMM-1 units (Adapted from Perleberg et al. (2020b))

much more 4×4 PUs than PUs with other sizes to be evaluated in a frame, the processing of 4×4 PUs is the bottleneck of the proposed architecture, defining the architecture throughput.

4.3.4   Synthesis Results and Comparison

The developed hardware architecture was described in VHDL and synthesized to ASIC with the TSMC 40 nm standard-cells library by using the Cadence RTL Compiler. The ASIC results show that the developed design can reach a maximum frequency of 1.8 GHz. At this frequency, the architecture is able to process up to nine views of 3D HD 1080p videos at 30 fps, while dissipating 263.7 mW. The developed design requires a cell area of 1,318,564 μm^2, or 1401 k gates (2-inputs NAND). When running at 1.2 GHz, it is able to process six views of a 3D HD1080p video in real-time, at 30 fps, while dissipating 172.5 mW. The necessary frequency to process only one view is 199.7 MHz. At this frequency, the developed hardware presents a power dissipation of 66.7 mW, which results in an energy-efficiency of 1.072 nJ/sample. The rest of the synthesis results are presented in Table 4.7, along with the related works results.

Table 4.7 Synthesis results

Parameters	(Sanchez et al. 2019)	(Amish et al. 2019)	Our 6WR design
Intra modes	DMM-1 and 4	DIS, DMM-1 and 4	DMM-1
BD-Rate (%) (random-access)	0.09	1.115[a]	1.162 (0.834[a])
Technology	ASIC@28 nm	FPGA@Xilinx Virtex 6 (40 nm)	ASIC@TSMC 40 nm
Area (k gates)	350.8	330[b]	1401
Memory (bits)	Yes (217.63k)	Yes (1.95 M)	No
Maximum throughput	HD1080p@30fps 1 view	HD1080p@30fps 6 views	HD1080p@30fps 9 views
Power for 1 view of HD1080p@30fps	151.6 mW	Not available	66.7 mW
Energy-efficiency 1 view of HD1080p@30fps	2.437 nJ/sample	Not available	1.072 nJ/sample

[a]Average results considering only four video sequences
[b]Estimated results considering only the LUT area

When comparing the obtained results with related works, the work in (Sanchez et al. 2019) shows a smaller impact on the encoding efficiency since the adopted algorithm is only based on removing the refinement stage. The design in (Sanchez et al. 2019) results in a lower gate count, however, it requires a dedicated memory of 217.63 k bits to store the wedgelets to be evaluated, and this additional area is not presented in its total gate count. Moreover, its maximum throughput is 62 Msamples/s, which allows the processing of one view of a 3D HD1080p video at 30 fps, while our design reaches the real-time processing of nine views from the same resolution. Besides, the results show that the developed hardware is 56% more energy-efficient than the hardware of (Sanchez et al. 2019) when operating at the same throughput (one view of HD1080p@30fps). One option to reach the processing of more than one view by (Sanchez et al. 2019) is to replicate its hardware, allowing the use of all features from the MVD format. However, this can highly increase its area requirements and power dissipation.

The impact in the encoding efficiency presented in (Amish et al. 2019) was measured using only a part of the recommended CTC video sequences (Müller et al. 2014). Only four of the eight recommended video sequences were used for only the RA profile. Thus, considering the accumulation of BD-Rate results (based on Table 4.6) for RA profile from the same four videos evaluated by (Amish et al. 2019), the impact in the encoding efficiency of the 6WR is reduced to 0.834%, which indicates a smaller impact on the coding efficiency than the one obtained by (Amish et al. 2019). Moreover, (Amish et al. 2019) was focused on FPGA technology, being the area results estimate in gate count.

The conversion from the ALM/LUT to gate count was performed according to (AMD 2024; Intel 2024), targeting a fairer comparison between these works. It was important to mention that this conversion considers only the area required by the LUTs, while it disregards the area of the other internal elements (communication, multipliers, memories, etc). Besides, (Amish et al. 2019) also requires a memory of 1950 M bits to store the wedgelets to be evaluated, which has a huge impact in the required hardware resources, while our design implements the Bresenham algorithm to avoid the memory requirement. Finally, the developed hardware has achieved similar throughput than (Amish et al. 2019), however (Amish et al. 2019) does not present its power results, so it is difficult to provide a fair comparison in terms of energy-efficiency.

4.4   Quality-Power Configurable Flexible Coding Order Hardware Design for Real-Time 3D-HEVC Intra Frame Prediction

Considering the CCO and Random-Access (RA) temporal profile (Müller et al. 2014), the intra-frame prediction demands 12.5% of the computational effort (Afonso et al. 2019b). We analyzed that the novel intra-frame prediction modes of 3D-HEVC represent 6.0% of the total encoding time, and they are only applied over depth maps. Therefore, the computational effort required by the novel 3D-HEVC intra-frame prediction tools evidences the need for the development of strategies with a focus on reducing complexity from the 3D-HEVC intra-frame prediction.

In this context, dedicated hardware designs focusing on video coding are mandatory and have been proposed by different works on literature (Lung et al. 2019; Alcocer et al. 2019; Shi et al. 2020; Huang et al. 2016; Afonso et al. 2017; He et al. 2015; Jou et al. 2015; Shi et al. 2021; Gogoi et al. 2021; Afonso et al. 2019a; Zhang et al. 2019; Sanchez et al. 2016; Amish et al. 2019; Ücker et al. 2020). Although some of those works present hardware solutions for a set of 3D-HEVC intra-frame modes, none of them have support for all intra-frame modes of the 3D-HEVC, which can process complete access units (BV and DVs including luminance, chrominance, and depth maps). Therefore, one can conclude that there was an important open opportunity to explore inter-channel redundancies between texture and depth map channels when considering complexity reduction heuristics for 3D videos. To explore this opportunity, this book presents a **complete 3D-HEVC intra-frame prediction system** with the main novel contributions presented below:

- A mode distribution and computational effort analysis of the intra-frame encoding tools and the block sizes considering FCO configuration.
- The development of hardware-oriented heuristics that explore inter-channel (between depth and texture) correlations using the FCO to reduce computational effort.

- A dedicated hardware design able to process all intra-frame prediction tools of the 3D-HEVC, for both texture and depth map, including luminance and chrominance.
- An architectural design able to reach real-time processing at 30 frames per second of three HD 1080p views with low power dissipation.

As previously mentioned, the 3D-HEVC divides each frame into 64 × 64 blocks named Coding Tree Units (CTU), which may be recursively partitioned into smaller blocks: Coding Units (CU) and Prediction Units (PU). CTU, CU, and PU acronyms indicate that different color channels are included, while Coding Block (CB) and Prediction Block (PB) acronyms are used when one color channel is considered alone.

For intra-frame prediction, the 3D-HEVC inherits the DC, Planar, and Angular modes from HEVC (Tech et al. 2016). Although these HEVC modes are efficient in encoding texture frames, they do not preserve depth map specific characteristics, as the presence of sharp edges with a large difference between two neighboring samples (usually in edges from the objects), and large portions of area with smooth value variation (inner parts of objects, which are in the same distance from the camera). Therefore, novel intra-frame prediction modes are adopted by 3D-HEVC to encode depth maps, such as the DMM-1, DMM-4, and DIS.

In the sequence, a statistical and computational effort analysis done to identify hardware-friendly constraints capable of reducing the complexity of the intra-frame encoding tools considering the FCO configuration is presented.

The experimental conditions used for the evaluations are based on the common test conditions made available by the JCT-3V group (Müller et al. 2014). Among these conditions, the document defines eight 3D-video sequences using MVD format, four different QP values to be used for texture (25, 30, 35, and 40), and four QPs for depth maps (34, 39, 42, and 45)(see Appendix A). The experiments used the 3D-HTM 16.0. The experiments consider the intra-frame prediction processing with RA temporal configuration (Müller et al. 2014) since this configuration considers a realistic operation scenario and to shows the relevance of intra-frame prediction tools even in situations where the 3D-HEVC inter-frames prediction tools are enabled. The statistical analysis considered the percentage of luminance samples encoded with each prediction mode and PU size.

Figure 4.11 shows the impact of each different intra-frame prediction tool in 3D-HTM. Figure 4.11a depicts the percentage of depth maps samples encoded using the DIS mode and the other intra-frame tools. Note that the DIS mode is extensively employed, encoding 86.68% of the intra-frame predicted depth map samples. Therefore, to better view and also to better evaluate the intra-frame predicted PBs according to the selected encoding tools, the other analyses disregard the DIS mode. Figure 4.11b, c present the representativeness of intra-frame prediction tools in depth map and texture frames, respectively. As can be noted, the 33 Angular modes encode a major part of intra-frame predicted samples on both

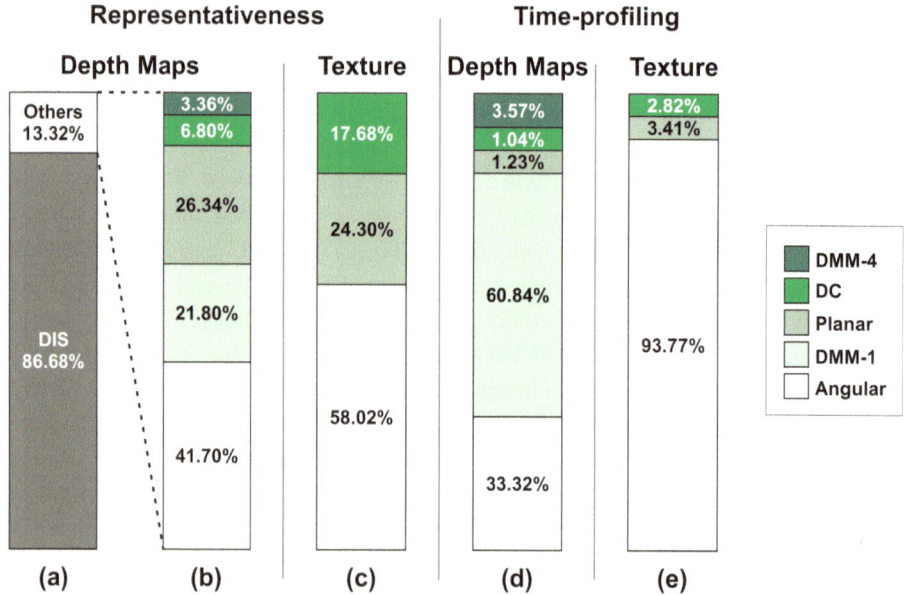

Fig. 4.11 FCO 3D-HTM intra-frame prediction tools representativeness and time-profiling analysis: **a** Representativeness in depth map samples considering DIS and other modes; **b** representativeness in depth map samples; **c** representativeness in texture frames; **d** time-profiling in depth maps; and **e** time-profiling in texture frames (Adapted from (Perleberg et al. (2022))

depth map and texture frames. Figure 4.11d, e present the computational effort related to each intra-frame encoding tool on depth maps and texture frames, respectively, where it can be observed that the Angular and DMM-1 modes have the highest complexity against the other intra-frame tools.

Figure 4.12 shows the analysis under the perspective of PU sizes. Figure 4.12a, b present the percentage of samples encoded with each PU size for depth map and texture frames, respectively, where it can be seen that the 4 × 4 PU size is less representative.

Figure 4.12c presents the computational effort of each PU size in depth maps. The 64 × 64 PU size demands the smallest computational effort mainly because the DMM modes (which are highly computationally intensive) are not applied to 64 × 64 PUs. It can also be observed that 4 × 4 PU size has a high complexity in depth maps. Finally, Fig. 4.122d presents the computational effort related to different PU sizes on texture frames, where one can conclude that 4 × 4 PU size is the most time-demanding in texture frames.

Therefore, to reduce the computational effort and to develop hardware-friendly heuristics, the main opportunities observed in the analysis and explored by the heuristics presented in the next section are:

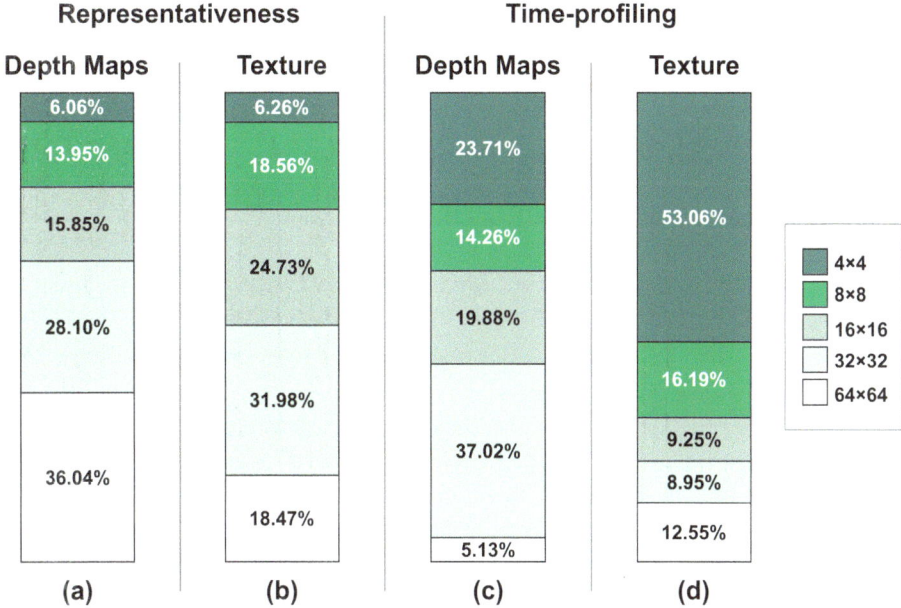

Fig. 4.12 FCO 3D-HTM different PU sizes representativeness and time-profiling analysis disregarding DIS: **a** Representativeness in depth map samples; **b** representativeness in texture frames; **c** time-profiling in depth maps; and **d** time-profiling in texture frames (Adapted from Perleberg et al. (2022))

- Avoid the support of 4×4 PU size due to its high complexity and low representativeness;
- Perform optimizations while exploring complexity reduction strategies for both Angular and DMM-1 modes due to their high complexities and high representativeness.

4.4.1   3D-HEVC Intra-Frame Prediction Hardware-Friendly Heuristics

The hardware-friendly heuristics developed along with this book exploit the 3D-HEVC behavior under FCO to better explore the correlations between texture and depth maps channels to reduce the required computational effort allowing an efficient hardware design. The following three subsections explain the three different hardware-friendly heuristics. In the sequence, the next section presents the impact on the coding efficiency of using the developed hardware-friendly heuristics.

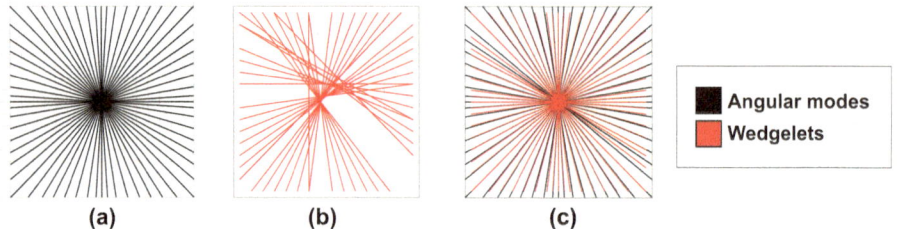

Fig. 4.13 Similarities between Angular and DMM-1 modes: **a** 33 angular modes; **b** 33 wedgelet patterns; **c** overlapping scheme with both the angular modes (red lines) and wedgelets (black lines) (Adapted from Perleberg et al. (2022))

4.4.1.1   Intra-Prediction Hardware-Oriented Constraints

The first constraint from Intra-Prediction Hardware-Oriented Constraints (IPHOC) implements hardware-friendly heuristics that impacts all intra-frame prediction encoding modes. It limits the supported PU sizes to the four most representative sizes considering the FCO approach, so that the developed design supports 8×8, 16×16, 32×32, and 64×64 PUs.

The second constraint from IPHOC was specifically defined considering the DMM-1 mode. As discussed before, the Angular modes support 33 different angles to predict directional structures. Since the DMM-1 mode was developed to efficiently predict regions containing sharp edges based on straight lines, it is possible to infer that there is a strong correlation between the decision of using Angular modes in texture channel and the decision of DMM-1 mode in the depth channel, when considering collocated PBs.

Therefore, we propose to reduce the total of 1908 possible wedgelets to use only 33 wedgelets for each PB size. Some directions employed in the DMM-1 wedgelets are the same used in the Angular modes while others are similar. So, for each PB size, it was selected the 33 wedgelets with directions most similar to the 33 Angular modes. Figure 4.13 show these similarities, where Fig. 4.13a shows the 33 Angular modes, Fig. 4.13b shows the 33 wedgelet patterns of equal or similar directions when compared to Angular modes, and Fig. 4.13c shows an overlapping scheme with both Angular modes (black lines) and wedgelets (red lines). Therefore, with the adoption of the IPHOC heuristics, no hardware resources should be dedicated to process neither the 4×4 PUs nor the remaining DMM-1 wedgelets, thus reducing memory accesses to obtain the wedgelet patterns and power dissipation from its processing.

4.4.1.2   Inter-Channel Directional Structure Detector

The regular flow applied to each PB in the 3D-HTM is presented in Fig. 4.14a, where it can be seen that it is generated the Most Probable Mode (MPM) list through the evaluation of all 35 HEVC-inherited intra-frame prediction modes. In a depth block, the DMM modes are also evaluated. From the MPM list, only a few modes are evaluated through

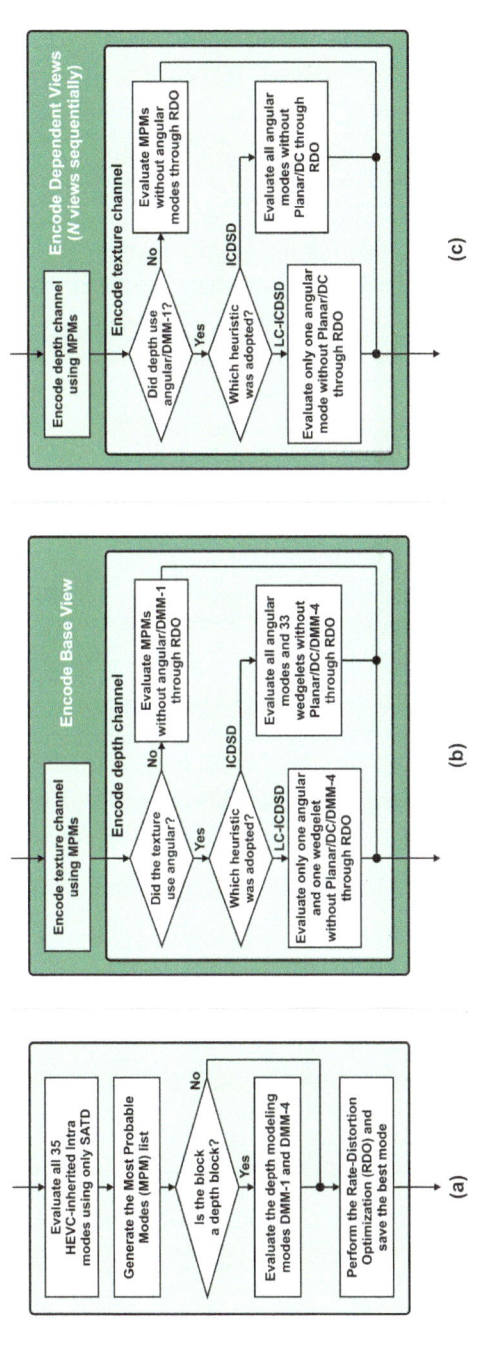

Fig. 4.14 3D-HEVC intra-frame prediction flowchart: **a** According to the 3D-HTM regular implementation; **b** encoding of BV according to our proposed ICDSD and LC-ICDSD heuristics; **c** encoding of DVs according to our proposed ICDSD and LC-ICDSD heuristics (Adapted from Perleberg et al. (2022))

the Rate-Distortion Optimization (RDO). In the FCO configuration, the luminance of texture information is used to aid the depth prediction in the BV as well as the use of depth information to aid texture prediction in the DVs (Gopalakrishna et al. 2013). Therefore, the Inter-Channel Directional Structure Detector (ICDSD) heuristic shares directional information between the two channels to define the intra-frame prediction modes to be evaluated through the RDO.

The ICDSD heuristic assumes that there is a correlation between collocated PB from two different channels. It is summarized for BV in Fig. 4.14b and for DV in Fig. 4.14c; also considering the LC-ICDSD described in the next subsection), which shows that the first channel from each view is normally encoded using the regular flow of 3D-HTM. However, in the second channel of each view, in the case where the collocated PB of the previously encoded channel does not present directional characteristics (i.e., the collocated PB does not use Angular or DMM-1), ICDSD avoids both Angular and DMM-1 evaluations. ICDSD benefits the hardware design by disabling the resources from Angular and DMM-1 modes when encoding the second channel if the first channel does not have directional characteristics, reducing power dissipation in those cases.

On the other hand, if the first channel has directional information, there is a chance that the second channel could also have directional information. So, if the collocated PB in the previous channel presents directional characteristics, the ICDSD skips the Planar, DC, and DMM-4 modes evaluation, as represented in Fig. 4.14b, c. However, in this case, the ICDSD evaluates all Angular and the 33 DMM-1 wedgelets (if depth map) through the complete RDO. This means that the Rough Mode Decision and MPM heuristics (Tech et al. 2016; Zao et al. 2011) are not used and, in these cases, a higher number of modes are evaluated than in the regular flow of 3D-HTM. On the hardware design, this means that the dedicated hardware resources for Planar, DC, and DMM-1 modes can be disabled when encoding the second channel, while all Angular modes will be enabled for evaluation. The DIS tool is not affected by ICDSD.

4.4.1.3 Low-Complexity Inter-Channel Directional Structure Detector

The Low-Complexity ICDSD (LC-ICDSD) heuristic also assumes the correlation in the directional information between collocated PB of two different channels. However, the LC-ICDSD uses the strategy of evaluating only one DMM-1 wedgelet and/or one Angular mode according to the decisions from the previously encoded channel. The BV is represented in Fig. 4.14b and shows that if the collocated texture PB has been predicted with an Angular mode and the LC-ICDSD is adopted, only the same Angular mode and one DMM-1 wedgelet with a similar direction are evaluated for the depth PB. Similarly, in the DVs, if the collocated depth PB has been predicted with Angular or DMM-1 modes, only the Angular mode with a similar direction is evaluated in the texture PU, as represented in Fig. 4.14c. Assuming that the hardware has 33 units dedicated to each Angular mode and 33 units to each DMM-1 wedgelet, then 32 Angular units and 32 DMM-1 units can be disabled to reduce power dissipation when encoding the second channel.

4.4.2   Hardware-Friendly Heuristics Coding Efficiency

The developed heuristics were organized in three distinct levels targeting a more efficient hardware implementation as the level increases: Level 0 adopts only IPHOC heuristic to restrict PU sizes and DMM-1 wedgelets. Level 1 adopts both IPHOC and ICDSD heuristics, thus increasing Level 0 by exploring the correlation between collocated PB of two different channels to disable Angular and DMM-1 modes. Level 2 adopts IPHOC, ICDSD, and also LC-ICDSD heuristics, thus it increases Level 1 by reducing, even more, the complexity since only one Angular and one DMM-1 wedgelet is evaluated, as previously explained.

The impact of using hardware-friendly heuristics is evaluated considering the impact on the coding efficiency based on the BD-Rate metric (Bjontegaard 2001) and the reduction in the computational effort based on the total encoding time. These heuristics were implemented in the 3D-HTM 16.2 and evaluated with the same test conditions described in Sect. 4.4 considering RA and All-Intra (AI) temporal configurations. The coding efficiency results are presented in Table 4.8.

In Table 4.8, the "Video Total" represents the coding efficiency considering the total bit rate of the video and quality measured over the encoded texture channels, i.e., it does not adopt a DIBR process to render new synthesized texture views. On the other side, "Synthesized Frames" presents the BD-Rate impact considering the total bit rate of the video and the quality measured over synthesized views using the information from both texture and depth maps by the DIBR, thus representing the impact on the quality of the views generated at the decoder side.

As presented in Table 4.8, the more restrictive the heuristic, the higher the impact on the coding efficiency. Level 0 presented a small BD-Rate impact, while it drastically saves hardware resources due to reduce the number of wedgelets evaluated in the DMM-1 and skips the 4×4 PU size, allowing a low-complexity hardware implementation. Level 1 enables ICDSD heuristic to disable the evaluation of Angular and DMM-1 modes when the previous channel does not present directional information, or increase the number of modes evaluated through the complete RDO (evaluating all Angular and DMM-1 modes) when the previous channel has presented directional information, thus resulting in a higher BD-Rate and encoding time when compared to Level 0. Finally, Level 2 results in the highest BD-Rate impact since it adopts LC-ICDSD heuristic to evaluate a small set of modes according to directional characteristics from the previous channel, enabling even more benefits for the hardware design.

In AI temporal configuration, only intra-frame prediction tools are enabled. Therefore, the obtained encoding time reduction is more expressive in AI than in RA configuration since the LC-ICDSD reduces up to 35 intra-frame prediction modes in the second channel being encoded to highly reduce the power dissipation from the hardware architecture.

In the CCO configuration, a complexity reduction by inheriting decisions between different channels is only possible in the depth maps since the texture is always encoded

Table 4.8 BD-rate variation by using the hardware-friendly heuristics in each evaluated sequence

Temporal Config.	Resolution	Sequence	Level 0 (IPHOC)			Level 1 (IPHOC + ICDSD)			Level 2 (IPHOC + ICDSD + LC-ICDSD)		
			Video total (BD-Rate)	Synthesized frames (BD-Rate)	Time reduction (%)	Video total (BD-Rate)	Synthesized frames (BD-Rate)	Time reduction (%)	Video total (BD-Rate)	Synthesized frames (BD-Rate)	Time reduction (%)
Random access	1024 × 768	Balloons	0.23	2.28	5.56	0.61	2.90	2.69	1.44	4.27	7.10
		Kendo	0.25	1.67	5.80	0.87	2.40	3.18	1.44	3.40	7.62
		Newspaper	−0.08	4.73	5.48	0.70	6.04	3.34	2.51	9.00	8.12
	1920 × 1088	GTFly	1.55	2.90	6.51	1.84	4.03	4.22	2.46	4.93	8.30
		PoznanHall2	0.57	2.16	6.29	1.15	3.01	2.96	1.89	4.38	5.13
		PoznanStreet	0.45	1.94	7.34	0.73	2.46	2.83	1.39	3.63	6.38
		UndoDancer	1.53	2.78	6.31	2.28	5.35	13.74	3.25	6.81	17.14
		Shark	0.49	2.16	6.27	1.62	5.00	2.39	2.47	6.39	7.40
Average RA			0.63	2.58	6.23	1.22	3.90	4.54	2.11	5.35	8.35
All Intra	1024 × 768	Balloons	0.63	6.11	47.55	4.41	10.25	28.05	22.28	28.24	61.77
		Kendo	0.58	5.96	48.12	4.89	10.74	29.74	27.13	33.05	61.68
		Newspaper	0.22	8.81	45.67	3.14	13.01	24.53	21.38	34.28	63.99
	1920 × 1088	GTFly	3.43	7.27	48.80	7.84	13.17	26.76	31.03	34.76	62.45
		PoznanHall2	0.71	6.14	49.52	10.79	17.48	25.16	32.70	41.15	61.70
		PoznanStreet	0.77	3.20	47.17	5.42	8.18	20.75	24.92	27.33	61.18
		UndoDancer	1.87	7.97	48.99	4.85	14.80	28.53	14.89	28.06	67.13
		Shark	0.44	7.54	49.22	3.95	14.40	25.44	15.43	26.58	61.80
Average AI			1.08	6.63	48.13	5.66	12.75	26.12	23.72	31.68	62.71

firstly. With FCO configuration, it is possible to reduce the complexity of the depth maps on BV, and in texture frames on DV by exploring decisions from the depth map of the same view.

As one can observe, the proposed heuristics result in an important increase in the BD-Rate but lead to expressive reductions in the encoding time. Also, these heuristics proved to have a high potential to be efficiently implemented in a hardware design since these heuristics only evaluate a small set of DMM-1 wedgelets while skipping the evaluation of 4×4 PU size. Moreover, they dynamically disable the evaluation of several intra-frame prediction modes according to the directional information inherited from the previous channel. All these features make these heuristics hardware-friendly, allowing high-throughput and energy-efficient hardware design, as described in the next section.

4.4.3   Quality-Power Aware Configurable 3D-HEVC Intra-Frame Prediction Hardware

The high-level view of the designed intra-frame prediction architecture is presented in Fig. 4.16a. This architecture implements the IPHOC hardware constraint, then the 4×4 PBs are not supported and the DMM-1 supports only 33 wedgelets directions to each PB size. Also, it can implement any Level, in a configurable way, where an external signal indicates to the Heuristic Control which heuristics will be active according to the desired energy savings, as discussed in Sect. 4.4.2.

The architecture contains four main units, where each one evaluates a subset of intra-frame prediction modes. As one can observe in Fig. 4.15a, each unit receives as inputs the edge samples from neighboring blocks for processing. After obtaining the predicted line from each mode, a multiplexer selects the predicted line of the intra-frame mode being processed (excepting the DMM-1 unit that processes only one intra-frame mode) and a subtractor computes the residue block. The next subsections describe these four units.

The application of ICDSD and LC-ICDSD heuristics are defined by the Heuristic Control unit that enables or disables the units presented in Fig. 4.15a by using the *HCEnable* and *HCAngle* signals. *HCEnable* is a 1-bit signal that indicates if the Multi-mode unit (Planar, DC, and DMM-4) is enabled or not. *HCAngle* is a 6-bit signal that indicates if the Directional and DMM-1 units are disabled (value 0), if all units are enabled (value 1) or if one specific unit is enabled (values 2 to 34). The hardware is disabled by using clock gating, which also disables specific parts of the units when processing smaller PBs, ensuring the power efficiency of the designed hardware.

A flowchart from the Heuristic Control is presented in Fig. 4.16, where it can be seen that when processing the first channel or adopting Level 0, all units are enabled. On the other hand, when ICDSD is active, the information inherited from the collocated PB in the previous channel defines which modes are enabled. If the collocated PB does not present directional behavior, then the *HCEnable* enables the Multi-mode unit, while *HCAngle* disables Directional and DMM-1 units. If inheriting directional behavior, then

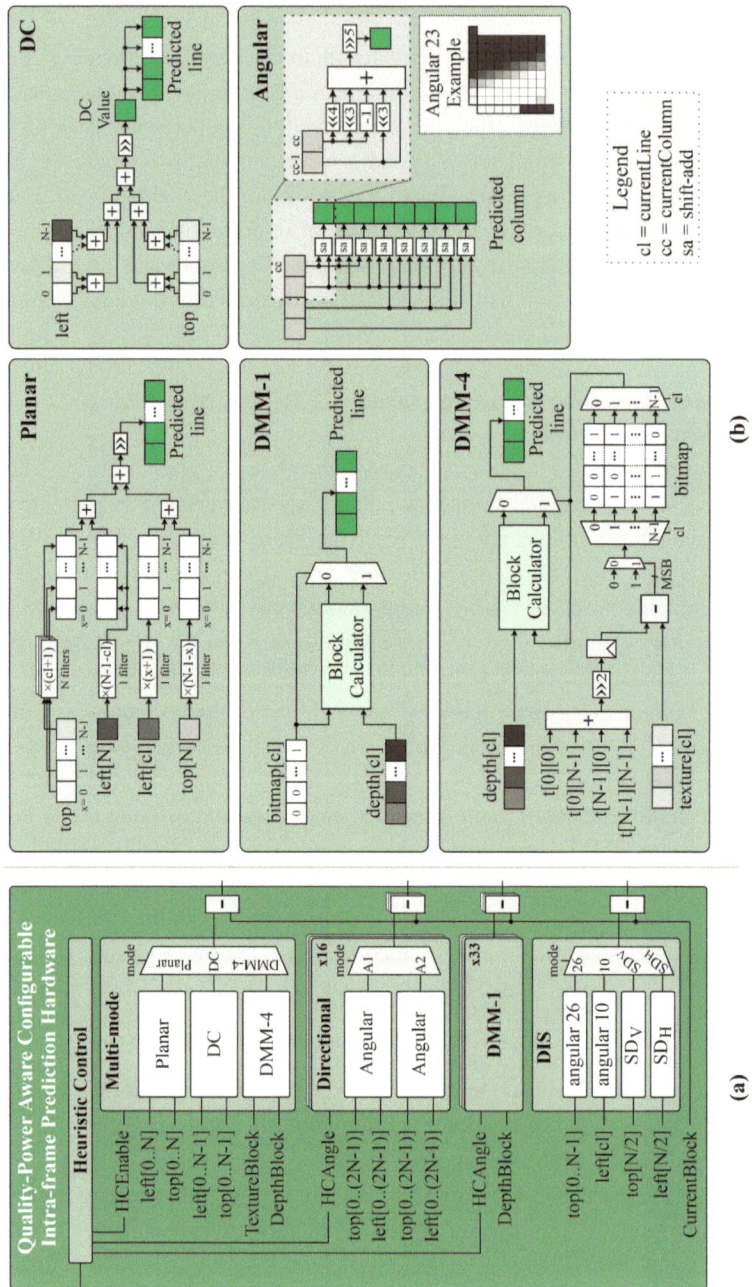

Fig. 4.15 Block diagram of the designed hardware: **a** High-level view; **b** planar, DC, DMM-4, angular and DMM-1 units (Adapted from Perleberg et al. (2022))

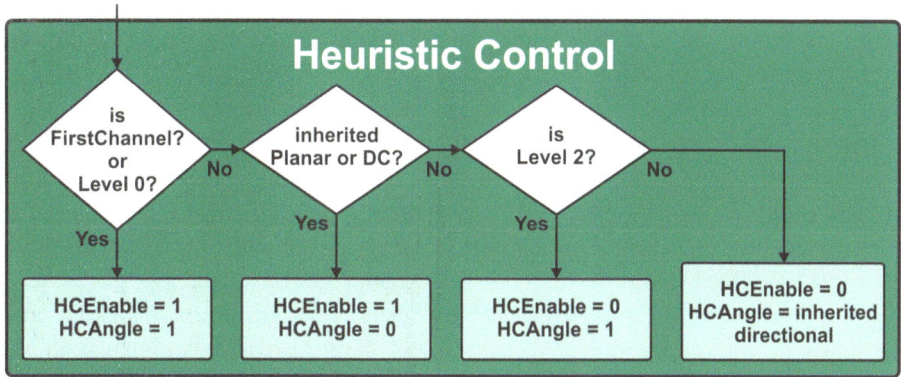

Fig. 4.16 Flowchart from the Heuristic Control detailing the decisions needed to control the architecture (Adapted from Perleberg et al. (2022))

the Multi-mode unit is disabled, while *HCAngle* enables both Directional and DMM-1 units. However, when the LC-ICDSD is active in Level 2, the *HCAngle* indicates which of the 33 angles was inherited from the previously encoded channel, enabling only one of the 33 instances of DMM-1 units (for depth channel) and only one of the 16 Angular units. The DIS unit is always enabled for depth maps since the developed heuristics do not affect its processing. Figure 4.15b presents each unit considering the processing of one N × N PB. For simplicity, the clock gating control used to enable/disable each unit is not presented in Fig. 4.15b. Each unit has parallelism enough to deliver a full line of the residue block at each clock cycle. This means the architecture can process texture PBs up to 64 × 64 samples, and depth PBs up to 32 × 32 samples. The DIS unit in Fig. 4.15a is not presented in Fig. 4.15b since it is quite simple, consisting of only copies from the neighboring inputs to the outputs.

4.4.3.1 Multi-mode Unit

The Multi-mode unit sequentially processes the predicted block of Planar, DC, and DMM-4 modes. These modes were grouped since the ICDSD heuristic applies the same restrictions to these modes. Figure 4.15b shows the processing unit from those modes.

In the Planar mode, the predicted block is computed from the weighted average of its four neighboring samples, located at the edge of the PB (Lainema et al. 2014). Each sample is predicted by multiplying the value of the edge sample by the distance between the predicted sample and the edge sample (Lainema et al. 2014). This distance can vary from 1 to 64, which means a multiplication by a constant value. So, the multipliers were replaced by shift-adder operations to reach higher throughput and to provide hardware resources and energy consumption savings. This way, a filter was developed with shift-adds that receives as input the edge sample, and results in the weighted value of up to 64 different positions. Since the processing is performed line by line, 67 filters are used to compute the predicted block.

The Planar unit is presented in Fig. 4.15b, where the values inside the filter operators show the position that each input sample is multiplied. As one can see, the Planar unit has 64 filters to process each top edge sample, being only one output from each of those filters used to compose the weighted average, according to the line being predicted. It also has one filter to process the lowest edge sample from the left neighbor, which output is replicated to complete the weighted vector. The Planar unit also has two other filters for processing the two edge samples that consider the sides of the line being predicted. After the filtering operations, each predicted sample is computed by accumulating the four related weighted values. To reach the throughput required to process high-resolution videos, a pipeline register is adopted after the filters. Thus, this Multi-mode unit needs $p(N)$ clock cycles to obtain the predicted block of the Planar mode considering an N × N PB, as given by (4.1).

For the DC mode, the predicted block is represented by the dcval, which is the average value of all neighboring samples from the processed PB (Lainema et al. 2014). Therefore, all edge samples from neighboring blocks are accumulated, two by two, as presented in the DC unit in Fig. 4.15b. After, a shift-right is applied according to the PB size to obtain the dcval. The Multi-mode unit requires $d(N)$ clock cycles to compute the predicted block of the DC mode considering an N × N PB, as given by (4.2).

From the collocated texture PB, the DMM-4 unit receives the values of the four corner samples (Ücker et al. 2020). As represented in Fig. 4.15b, firstly, the average of these four corner samples is obtained. Based on this average, the bitmap block representing the contour regions is computed, line by line. Finally, the Block Calculator processes this bitmap block along with the current depth PB, to compute the average value from each region of the predicted block. The DMM-4 unit has four pipeline registers, being one located after obtaining the average of the four texture samples, and three inside the Block Calculator. The DMM-4 unit requires $v(N)$ clock cycles to compute the predicted block of the DMM-4 mode considering an N × N PB, as given by (4.3).

As previously mentioned, the Multi-mode unit sequentially processes its three intra-frame prediction modes. This means that one intra-frame mode is processed while the rest of the unit is disabled using clock gating, reducing power dissipation. This approach results in $u(N)$ clock cycles to process the three modes of this unit for each PB, as given by (4.4).

$$p(N) = N + 1 \tag{4.1}$$

$$d(N) = N \tag{4.2}$$

$$v(N) = 2N + 4 \tag{4.3}$$

$$u(N) = p(N) + d(N) + v(N) \tag{4.4}$$

4.4.3.2   Directional Unit

The Directional unit is responsible for processing only the Angular modes. As it will be presented, the DIS unit processes the Angular modes 10 and 26. So, the Directional unit processes the other 31 Angular modes. The computation of these 31 modes is divided into 16 instances of the Directional Unit, where each one has two Angular Units responsible for processing two Angular modes. All these units operate in parallel, filled by the same neighboring input samples.

When the ICDSD is active, all 16 instances of the Directional Unit can be enabled or disabled through the *HCAngle* signal provided by the Heuristic Control unit. When the LC-ICDSD is active, the *HCAngle* indicates the direction inherited from the previous channel, and only one Angular Unit is enabled while the other 30 Angular Units are disabled.

To process each mode, each sample from the predicted block is generated based on the weighted average of some specific neighboring samples, according to the direction of the processed Angular mode. For that, each Angular unit adopts the strategy of deriving a reference vector from the neighboring samples before predicting the samples of the PB using that vector (Lainema et al. 2014). This enables the sharing of the weights between all the lines or between all the columns from the predicted block.

Figure 4.15b exemplifies the prediction of the intra-frame mode 23, presenting the four input samples necessary to process the eight first samples of the current column being predicted. As one can notice, each sample from the column being predicted is computed by the weighted average of two neighboring samples. So, to process one complete column (or line (Lainema et al. 2014)), the hardware requires all resources to multiply the specific neighboring samples by the weight values. These weight values were fixed to all columns. So, the multiplications were implemented with shift-adds operations, reducing the cost of these operations. Since each Directional Unit processes the residue of the two Angular modes sequentially, the processing of the Directional Unit to an N × N PB requires $a(N)$ clock cycles, as given by (4.5).

$$a(N) = 2 * N \qquad (4.5)$$

4.4.3.3   DMM-1 Unit

The designed architecture for the DMM-1 has 33 DMM-1 units working in parallel to obtain the predicted block of the 33 wedgelets to be evaluated. Each DMM-1 unit has an internal Read-Only Memory (ROM) storing its related bitmap, which is delivered one line per clock cycle. The *HCAngle* signal works similar than in the Directional Unit, where the ICDSD heuristic can disable all DMM-1 Units, while the LC-ICDSD heuristic indicates which DMM-1 Unit should be enabled.

Figure 4.15b presents the architecture that processes one wedgelet for the DMM-1 mode. This architecture receives a bitmap line (from its internal memory) representing the PB division for a specific wedgelet. Then the bitmap is forwarded to a Block Calculator unit, where the average value from each region is generated. Thus, each sample of the predicted block receives the average value of its respective region. The Block Calculator performs its processing line by line, and it has four pipeline registers.

Since the DMM-1 Unit is used only in a small number of cycles compared with the other units, the local clock used in Block Calculator is divided by two. This technique reduces the dynamic power related to the average calculation for the two regions of the predicted block. Therefore, this unit needs $w(N)$ global clock cycles to process an N × N PB, as given by (4.6).

$$w(N) = 3N + 8 \tag{4.6}$$

4.4.3.4 DIS Unit

Finally, the DIS unit is responsible for processing the four DIS sub-modes and the Angular modes 10 (horizontal) and 26 (vertical). This strategy is used since the processing performed for DIS sub-modes and these two Angular modes is the same. Since the proposed heuristics do not affect DIS modes, the DIS unit is always available for processing, being only disabled using clock gating when it has finished the processing.

To obtain the predicted block of each DIS sub-mode, the values from neighboring samples are copied to the predicted line, without modifying its original value (Afonso et al. 2017). Similar computing is performed for mode 26 (vertical Angular or DIS IPV mode), where the predicted line receives a copy of the value from the neighboring sample above the line being processed. Processing any DIS sub-mode to an N × N PB requires $k(N)$ clock cycles, as given by (4.7).

$$k(N) = N \tag{4.7}$$

4.4.3.5 Architecture Synchronism and Memory Bandwidth

Figure 4.17 shows a timing diagram considering all four units when processing an N × N PB. It considers that all four units are enabled, but some units may be disabled according to ICDSD and LC-ICDSD heuristics, PB size and encoded channel. The Multi-mode is generally the unit that requires a higher number of clock cycles. It starts by processing the predicted block of the Planar mode followed by the DC mode. In the case of a depth map PB, the DMM-4 is processed after the DC mode. Directional and DIS units also process each mode sequentially.

The hardware starts processing the BV, where the luminance and chrominance from the texture are encoded first by Multi-mode, Directional, and DIS units, while the other units are disabled (used only for depth maps) using clock gating to reduce power dissipation. After, the depth map is encoded by the units defined by the adopted heuristic. When the DVs are processed, the depth map of each DV is encoded first using all units, and after

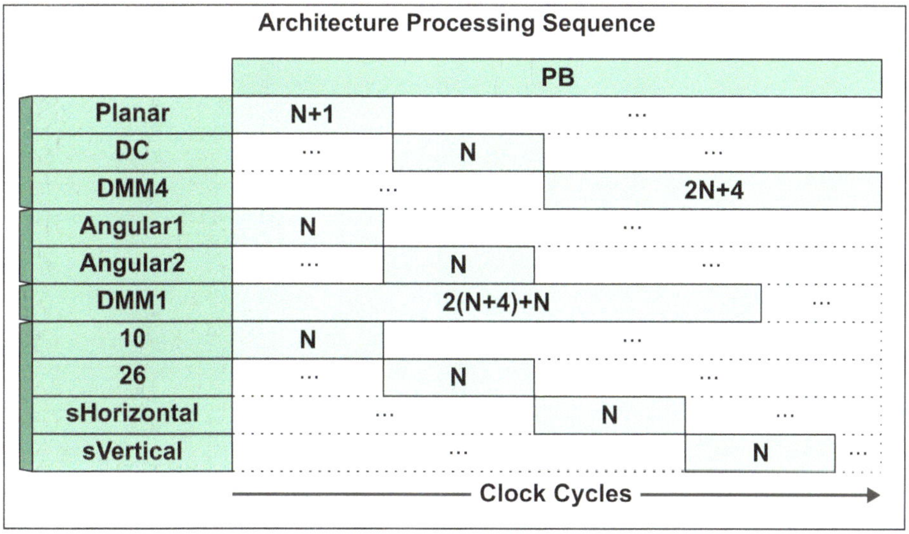

Fig. 4.17 Timing diagram of the proposed architecture considering one $N \times N$ PB (Adapted from Perleberg et al. (2022))

the luminance and chrominance of each DV texture frame are encoded using Multi-mode (Planar and DC), Directional, and DIS units, according to the adopted heuristic.

At the level of CTU encoding, the designed hardware sequentially processes the four PB sizes from each CTU of the frame being processed. The hardware units were designed to process the 64×64 maximum PB size (the DMM units support the maximum PB size of 32×32 samples). The architecture starts processing the CTU as a 64×64 PB, and after it splits the CTU to process the smaller PBs. The parts of the units that are not necessary to process the smaller PBs are also clock gated.

Table 4.9 shows the number of clock cycles required by the proposed hardware to process each N × N PB. In the proposed Level 1 and Level 2, some units are disabled by exploiting the directional information from the collocated PB. In comparison with Level 0, Level-1 only affects the number of clock cycles in the second channel of each view. However, since all four units from the architecture run in parallel, only a small difference in the number of cycles was observed in cases where the collocated PB has directional information.

The "Full CTU" row considers the processing of the four supported PB sizes from the four collocated CTUs (three texture channels and the collocated depth map). Table 4.9 also presents the total number of cycles required to process all PBs from the four collocated 64×64 CTUs of each view considering a conventional access unit with three views (the BV and two DV; Müller et al. 2014). As can be seen, the proposed hardware requires at least 30,825 clock cycles to process all collocated 64×64 CTUs from three views.

Table 4.9 Number of required clock cycles to process an N × N PB

View	Channel	PB size	Clock cycles to each heuristic						Without IPHOC Heuristic
			Level 0	Level 1		Level 2			
				Without directional info	With directional info	Without directional info	With directional info		
Base view	Texture	Any	2N + 1	2N + 1	2N + 1	2N + 1	2N + 1		2N + 1
	Depth map	64	4N	4N	4N	4N	4N		4N
		32	4N + 5	4N + 5	4N	4N + 5	4N		48N + 128
		16	4N + 5	4N + 5	4N	4N + 5	4N		48N + 128
		8	4N + 5	4N + 5	4N	4N + 5	4N		75N + 200
		4	–	–	–	–	–		9N + 24
	Full CTU		10,275	10,275	9,855	10,275	9855		100,735
Dependent view	Depth map	64	4N	4N	4N	4N	4N		4N
		32	4N + 5	4N + 5	4N + 5	4N + 5	4N + 5		48N + 128
		16	4N + 5	4N + 5	4N + 5	4N + 5	4N + 5		48N + 128
		8	4N + 5	4N + 5	4N + 5	4N + 5	4N + 5		75N + 200
		4	–	–	–	–	–		9N + 24
	Texture	Any	2N + 1	2N + 1	2N	2N + 1	N		2N + 1
	Full CTU		10,275	10,275	10,020	10,275	7140		100,735
All collocated CTUs from three views			30,825	30,825	29,895	30,825	24,135		302,205

This value is reduced by 3.02 and 21.70% when using Level 1 and Level 2, respectively, when the collocated PB has directional information. The number of cycles from Level 1 and Level 2 is the same as from Level 0 in the cases where the collocated PB does not have directional information since the Multi-mode unit requires more cycles than other units. When the collocated PB has directional information, it is the DIS unit (in the depth map of BV) or Angular unit (in the Texture of DV) that requires more clock cycles. So, in Level 2, the reduction is higher than in Level 1 since only one Angular mode can be evaluated in the texture of DV when the collocated PB has directional information.

Finally, for the sake of comparison, Table 4.9 also presents an estimated number of clock cycles from a hypothetical case where the developed architecture does not implement the IPHOC heuristics, thus evaluating the 4×4 PB size and all 1908 DMM-1 wedgelets. In this hypothetical case, only the 33 DMM-1 units from the proposed architecture were adopted to process all the 1908 original wedgelets (Zhang et al. 2014; Ikai et al. 2015) by repeating their processing several times. Also, the proposed architecture processes the 4×4 PB size after the processing of all other PB sizes. From this hypothetical case, one can observe that without the IPHOC constraints, the number of clock cycles required to process all CTUs by the architecture would increase by more than 9.8 times. Consequently, operating frequency and energy consumption would be highly increased.

Therefore, considering the clock cycles to process all collocated 64×64 CTUs presented in Table 4.9, and considering the processing of an HD 1080p video at 30 fps, which has 510 CTUs per frame, the developed hardware requires an operating frequency of 471.6 MHz, assuming the worst case where all collocated PBs do not have directional information. Since the number of clock cycles to process the worst case does not suffer variation no matter which Level is adopted by the developed hardware, the operational frequency of 471.6 MHz is also the same for any adopted Level. This means that at 471.6 MHz, the designed hardware can implement any Level, processing the 3D-HEVC intra-frame prediction with three views of HD 1080p video at 30 fps, including texture and depth maps channels and including luminance and chrominance samples.

The proposed hardware requires access to both texture block samples and their collocated depth map block samples, and also the neighboring samples of those blocks. It also requires one extra Byte to store which prediction mode was used in the first channel since the proposed hardware does not internally store this information. Therefore, our Intra-frame system requires an SRAM of 8449 Bytes. Considering the parallel processing of all intra-frame modes with the same input samples, the designed hardware requires up to 7936 new input samples to process all PB sizes of a 64×64 texture CB and up to 12,032 new input samples to process all PB sizes of a 64×64 depth map CB. So, considering the processing of three views of an HD 1080p video at 30 fps, our hardware requires 1645.06 M samples per second, requiring a bandwidth to the SRAM of 1645.06 MB/s when 8-bit samples are considered.

4.4.4 Results and Comparison

The architecture was described in VHDL and synthesized to ASIC using Cadence RTL Compiler tool and TSMC 40 nm standard-cells library (TSMC 2024; Perleberg et al. 2018). The ASIC results show the architecture can reach the maximum frequency of 786.1 MHz, being able to process up to five HD 1080p views at 30 fps. Table 4.10 shows the developed hardware requires 7933 k gates, being 68.3% required by the 16 Directional Units and 19.9% dedicated to the 33 DMM-1 Units. Other 8.5 and 3.2% are dedicated to the Multi-mode and DIS units, respectively.

Since the developed hardware focuses on processing access units with three views at HD 1080p video resolution, we present the power results for 471.6 MHz. Also, we present power results at the maximum frequency of 786.1 MHz, required for processing five views of HD 1080p videos at 30fps. Power results consider statistical switching activity (Perleberg et al. 2018) considering the representativeness of Angular and DMM-1 modes presented in Fig. 4.11, thus it considers that 60% of the CTUs have directional information. This switching activity was adopted to control the clock-gating logic of the circuit and to simulate the disabling of specific units according to ICDSD and LC-ICDSD heuristics, estimating the power dissipation from the three different Levels of the heuristics. According to Table 4.11, when processing three HD 1080p views videos at 30 fps, the proposed hardware dissipates 528.8, 473.3, or 384.6 mW, depending on the adopted Level. When processing five views, the power dissipation ranges from 749.4 to 493.5 mW. The synthesis results show that the static power is 186 mW, indicating that the dynamic power of Level 2 in comparison with Level 0 suffers a reduction of 42.2 and 45.7% when processing three or five views of HD 1080p videos at 30 fps, respectively.

Table 4.10 Cell area distribution among hardware units

	Multi-mode unit	Angular unit	DMM-1 unit	DIS unit	Total
Cell area (gates)	674,318	5,422,803	1,579,973	256,280	7,933,376
	8.50%	68.35%	19.92%	3.23%	

Table 4.11 Power dissipation when processing HD 1080p videos at 30fps

Views in access unit	Power (mW)		
	Level 0 (IPHOC)	Level 1 (IPHOC + ICDSD)	Level 2 (IPHOC + ICDSD + LC-ICDSD)
3 views (471.6 MHz)	528.8	473.3	384.6
5 views (786.1 MHz)	749.4	639.2	493.5

Taking into account the BD-Rate results (presented in Table 4.8 in Sect. 4.4.2) and the power results from Table 4.11, it can be seen that Level 0 presents the highest power dissipation but provides the lowest impact in coding efficiency since none of the ICDSD/LC-ICDSD heuristics is enabled to reduce the number of intra-frame evaluated prediction modes. By Table 4.11 it can also be seen that Level 1 has resulted in smaller power dissipation when compared to Level 0. This reduction is related to the distribution of directional modes in the first channel encoded and changes according to the video characteristics. With ICDSD heuristics, when the collocated PB has directional information, all Angular and DMM-1 Units (the most complex units and responsible for large area requirements) are enabled and the power reduction is limited. In turn, Level 2 have presented the highest impact in the coding efficiency, but it proved to be a good option for battery-powered devices, since with LC-ICDSD heuristics, when the collocated PB has directional information, only one Angular and one DMM-1 Unit are enabled, resulting in a power reduction ranging from 18.75 to 22.8% compared to Level 1.

The IPHOC evaluating only 33 DMM-1 wedgelets to each PB size proved to be a good hardware-friendly solution since significant extra hardware would be needed to evaluate all 1908 DMM-1 wedgelets sustaining performance. Without our heuristic, using the available 33 DMM-1 units to process all 1908 DMM-1 wedgelets would require a frequency of 4.62 GHz (maximum frequency is 786.1 MHz, as previously mentioned), leading to excessively high-power dissipation.

The adoption of FCO with proposed heuristics also proved to be a good hardware-friendly solution since the proposed heuristics skip the evaluation of up to 35 intra-frame prediction modes in DV, reducing complexity in both luminance and chrominance samples of the texture. If adopting the CCO, the first channel has all intra-frame prediction modes enabled to encode luminance and chrominance samples of the texture, while in the second channel, only the depth map would be affected by the heuristics disabling some modes.

To the best of our knowledge, the hardware design presented in this Section was the first in the literature proposing a complete hardware design for all intra-frame modes of the 3D-HEVC, which can process complete access units (BV and DVs including luminance, chrominance, and depth maps). However, there are a few works proposing dedicated hardware for a set of intra-frame modes and specific channels, such as depth maps or textures. Since these works do not have support for processing a complete access unit, a direct comparison in terms of compression efficiency results is unfair.

The obtained compression efficiency losses are expressive if directly compared with works proposing hardware solutions only to a set of video encoding tools. Although, in those works, the remaining tools (not implemented) are considered original, thus they reach smaller losses in compression efficiency. Moreover, considering that the "Video Total" column from Table 4.8 represents the BD-Rate impact in the encoded texture channels, the obtained results can be considered similar to the BD-Rate impact of other works presented in the literature proposing hardware designs for 2D video encoder as (He et al. 2015; 2.07%), (Jou et al. 2015 (4.67%), (Shi et al. 2021; 3.02%) and (Gogoi et al. 2021; 2.87%). Further, when comparing the "Synthesized Frames" results for RA from Table 4.8 with (Afonso et al. 2019a), which presents a dedicated hardware for the complete 3D-HEVC Motion and Disparity Estimation, it can be noted that we obtained a considerable small impact (5.35%) since that (Afonso et al. 2019a) have obtained a BD-Rate of 23.22%.

The characteristics and results from these works proposing dedicated hardware for specific intra-frame modes were presented in Table 4.12, along with results of our hardware architecture operating at Level 2. From Table 4.12 it can be seen that none of those related works have support for both texture and depth map, neither support for both luminance and chrominance samples. Also, none of those works propose an approach that explores FCO. Those works propose low-complexity solutions to only a few intra-frame prediction modes and, therefore, direct comparisons are not fair.

Table 4.12 Results of related works for dedicated hardware design to intra-frame prediction modes

Related works		(Zhang et al. 2019) TCSVT	(Sanchez et al. 2016) JRTIP	(Amish et al. 2019) JRTIP	Developed design
Texture plus depth support		No (only texture)	No (only depth)	No (only depth)	Yes
Luminance and chrominance support		Yes	No	No	Yes
Flexible coding order approach		No	No	No	Yes
Number of intra modes		35	2	3	38
Intra-frame prediction modes		Planar, DC, and 33 Angular	DMM-1, DMM-4	DIS, DMM-1, DMM-4	All Intra-Frame Modes from 3D-HEVC
BD-rate for synthesized frames (%)		RA: not Available	RA: 0.09	RA: 1.12	RA: 5.35
Number of DMM-1 wedgelets evaluated		–	1140	1	33
Technology		ASIC TSMC 90 nm	ASIC 65 nm	Xilinx FPGA Virtex 6	ASIC TSMC 40 nm
Total area		2288 K	219.95 k gates	55 k LUT, 67 k REG	7933 k gates
Power for 1080p@30fps	Frequency (MHz)	80 MHz	53.2–632.9	275	786.1
	Views	1	1	6	5
	Power (mW)	236	412.7	Not Available	493.5

Zhang et al. (2019) proposes hardware architecture for the HEVC intra-frame prediction modes. It has support to all 35 intra-frame prediction modes of the HEVC, but only up to seven were evaluated to each CU through the complete RDO. Since its architecture is focused on HEVC, depth maps and FCO are not considered. Therefore, its requirements were smaller than ours.

In the 3D-HEVC scope, most of the related works perform simplifications on the DMM-1 algorithm. Sanchez et al. (2016) presents architecture only for DMM-1 and DMM-4 modes, thus requiring a small area compared with our design. However, memory issues are not covered by this work. Also, its proposed algorithm was simplified by only removing the refinement process, thus reaching a small impact on the coding efficiency. Therefore, it was able to process only one 1080p@30fps view while requiring a similar power dissipation than our solution processing five 1080p@30fps views.

Amish et al. (2019) has proposed a hardware architecture for 3D-HEVC depth map specific tools considering all possible PU sizes. The strategy of its algorithm evaluates only one wedgelet for each PU size, being this wedgelet based on the gradient of the edges of the PU. Its architecture was focused on FPGA technology, which makes it difficult the comparison of area results. However, since (Amish et al. 2019) proposes a solution only for a few modes, its area requirements are obviously smaller than ours. The solution of (Amish et al. 2019) has reached a higher processing rate, but it does not present any power results to be compared. Further, memory aspects were not considered.

In addition, it is important to mention that a huge part (68.3%) of the hardware resources usage is due to the processing of Angular modes, which performs several multiplications, and none of the related works has support to process all Angular modes in parallel. Note that for these works to support all features our work does (texture, depth map, luminance, chrominance, and FCO), their architectures would demand an important increase in hardware resources.

Inter-Frames and Inter-View Predictions Developed Architectures

This chapter presents in detail the two high-throughput architectures developed for the Inter-frames and Inter-view predictions. The first architecture consists of a Motion/ Disparity Estimation system with run-time adaptive memory hierarchy that proposes several complexity-reduction heuristics to reduce energy and memory requirements. The second hardware design consists of a coding-efficient Disparity Estimation architecture based on the Improved Unidirectional Disparity-Search Algorithm (iUDS). It is important to mention that the complexity-reduction heuristics proposed for these architectures considers different contexts. The first architecture, i.e., the ME/DE system with run-time adaptive memory hierarchy focuses on an FCO approach in order to take advantage of the channel coding order to save energy, and, therefore, its heuristics are also proposed and evaluated in terms of compression based on the statistical results for the FCO approach in Chap. 3. In the second proposed architecture, i.e., the DE architecture based on the iUDS Algorithm, the focus is on the coding efficiency by considering all PU block sizes. Therefore, the impacts on the compression of the second architecture consider the CCO approach. However, it can be used along with an FCO approach without drastic modifications.

As well as in the Intra-frame developed architectures, both architectures developed for the Inter-frames and Inter-View predictions used pipeline and the clock-gating low-power technique in order to reach processing rates capable of processing high and ultra-high definition videos and, also, to save energy consumption when it is possible.

© The Author(s), under exclusive license to Springer Nature Switzerland AG 2025
V. Afonso et al., *Hardware Design for 3D Video Coding*, Synthesis Lectures on Engineering, Science, and Technology, https://doi.org/10.1007/978-3-031-80232-4_5

5.1   Energy-Aware Motion and Disparity Estimation System with Run-Time Adaptive Memory Hierarchy

Given the processing/memory-related challenges posed by ME and DE, several memory and energy-aware hardware designs for ME and DE targeting previous video coding standards have been published over the years. Some proposals focus on optimizing the processing-related issues by reducing the computational effort (Kim et al. 2014) or designing energy-efficient hardware architectures (Ding et al. 2010). Other works propose solutions aiming at efficient memory organization, sizing, and management (Zatt et al. 2011a; Sampaio et al. 2013; Song et al. 2015). Finally, a set of works proposes complete ME and/or DE systems by jointly discussing memory and processing-unit designs (Zatt et al. 2011b). There are many insightful published hardware solutions focusing on energy efficiency, i.e., taking into account processing and memory issues. However, there is only one publication based on these approaches for 3D-HEVC (Afonso et al. 2019a).

Actually, most of the 3D-HEVC-related papers regard algorithmic solutions only, and several of them present solutions already implemented in the 3D-HEVC Reference Software. ME and DE are applied to both texture and depth-map channels in 3D-HEVC. However, there are only one ME/DE hardware design capable of handling MVD format in the current literature. Since these channels present completely different characteristics, it may be beneficial to choose the channel to be first processed. 3D-HEVC enables this feature through the Flexible Coding Order (FCO) tool (Gopalakrishna et al. 2013) that allows the encoding of depth maps before their associated texture frames, except for the Base View. In the work (Afonso et al. 2019a), the FCO is exploited regarding memory and processing issues to reduce energy consumption.

Therefore, this book presents a 3D-HEVC Motion and Disparity Estimation system designed for low-energy consumption. This system features a dedicated memory hierarchy capable of run-time adaptation. The processing unit employs FCO along with a series of simplifications and optimizations targeting at computational effort reduction through data behavior observations and inter-channel/inter-view redundancies exploration. Statistical analysis was performed and motivated the proposal of an efficient average-case hardware design. Among the main contributions are the proposed of the Hardware-Oriented Test Zone Search (HOTZS) with static scheduling, which is a hardware-friendly version of the conventional HEVC/3D-HEVC fast motion estimation algorithm. It was proposed to allow the design of a high-throughput ME architecture. Also, the Horizontal Disparity Search (HDS) technique is proposed to avoid vertical search based on the horizontal-only camera displacement and propitiate a simplified and efficient disparity-estimation hardware design. The proposed system employs distributed on-chip memories associated with the processing units. Each memory features window-based data reuse and is composed of multiple sectors enabling independent control via Dynamic Voltage Scaling (DVS). Sub-sampling is applied to depth-map memories whereas Horizontal Disparity Search (HDS) allows memory size reduction. A Depth-Based Dynamic Search Window Resizing

(DSWR) algorithm is proposed to dynamically reduce the search window during texture ME/DE based on depth information, allowing dynamic power management through DVS of idle memory regions. The Texture-Based Motion Vector Inheritance (TMVI) technique takes advantage of the ME/DE behavior and inter-channel correlation to reduce energy consumption. The next subsection presents the definition of a Baseline Encoder (B-Encoder) configuration used throughout this system development based on extensive tests performed using the 3D-HTM reference software.

5.1.1   3D-HEVC Baseline Encoder Definition

As previously discussed, the intensive computation and the large amount of data handled by ME and DE lead to a large amount of memory access and high energy consumption in 3D-HEVC. These tools must be simplified by applying constraints/modifications to their processes in order to make the system implementation feasible. This means there is a need to define a set of hardware-oriented constraints focusing on complexity and communication reductions. However, applying such constraints usually impact negatively on the compression rates and image quality. Therefore, extensive tests were performed to evaluate the impact of different constraints on the 3D-HEVC encoder. This evaluation allowed the definition of a B-Encoder configuration used throughout this system development and evaluation.

The experimental setup used to define the B-Encoder configuration, and an analysis that led to the definition of the ME/DE hardware-oriented constraints were performed with the 3D-HTM in version 16.2 (3D-HEVC Reference Software 2024). The experiments follow the 3D-HEVC CTC document, recommended by the JCT-3V (see Appendix A). Therefore, the eight 3D video sequences (with three texture plus depth views), including five 1920×1088 (HD 1080p) sequences and three 1024×768 sequences (HD 760p) were used under the four recommended sets of QPs to encode texture/depth: 25/34, 30/39, 35/42, and 40/45. Also, the experiments considered the RA temporal configuration with 48 AUs, the three views defined by the CTC, and the Search Range (SR) of 64 pixels.

To define the ME/DE hardware-oriented constraints (HC), an incremental analysis was performed to verify the impact of them on the coding efficiency. After an extensive analysis of each HC considering different scenarios, one of these scenarios was adopted as the baseline for the next HC evaluation, and so forth. Each one of the ME/DE HCs is discussed in the following, and all results are summarized at the end of this section. The coding efficiency degradation is measured using the BD-Rate metric, which represents the percentage variation in bit rate considering the same objective image quality in comparison to the original 3D-HTM encoder. The coding efficiency results are discussed for video total (i.e., the quality related to texture channel and the total bit rate of the video) and for synthesized frames (i.e., the quality of frames synthesized with texture and depth using DIBR).

5.1.1.1   Block-Size Constraint (HC1)

To develop the ME/DE system, the first evaluation was performed to choose a subset of block sizes (known as PU—Prediction Unit) to be used in ME/DE process and avoid the evaluation of all 24 sizes possible in 3D-HEVC. In this hardware-constraint definition, the possible sizes were reduced to two. The choice of two PU sizes among 24 possibilities requires a proper analysis to identify the ones that yield a better tradeoff between complexity reduction and bit-rate increase. Since 276 combinations of two PU sizes are possible considering 24 PU sizes, a complete analysis of this space using the 3D-HTM is unfeasible. Given that the four square-shaped blocks (8×8, 16×16, 32×32, and 64×64) are the most frequently used and representative sizes in 3D-HEVC (see Sect. 3.2), the best combination of two square-shaped blocks (six possible combinations) was evaluated. The results regarding BD-Rate for these six combinations are presented in Table 5.1. Note that two test cases present a BD-Rate increase lower than 7% considering synthesized views. These two promising test cases consider only 16×16 and 32×32 PU sizes, or only 16×16 and 64×64 PU sizes in ME/DE and present very similar results. Thus, based on the BD-Rate results obtained, the 16×16 and 32×32 block sizes were adopted as the ME/DE block-size constraint since this approach allows the development of more efficient hardware solutions considering processing and memory-related issues.

The complete BD-Rate results, according to the 3D-video sequences, can be observed in Appendix C.

Table 5.1 BD-Rate increase according to the block-size constraint (HC1)

Block-size constraint (HC1)	Type of pictures	Average BD-rate increase (%)		
		1024×768	1920×1088	All sequences
8×8 and 16×16	Video total	9.07	15.12	12.85
	Synth	8.44	14.07	11.96
8×8 and 32×32	Video total	5.44	10.17	8.39
	Synth	5.12	9.43	7.81
8×8 and 64×64	Video total	7.56	13.42	11.22
	Synth	7.08	12.55	10.49
16×16 and 32×32	Video total	3.81	8.92	7.00
	Synth	4.19	8.44	6.85
16×16 and 64×64	Video total	4.38	8.54	6.98
	Synth	4.58	8.10	6.78
32×32 and 64×64	Video total	5.30	14.11	10.81
	Synth	5.92	13.64	10.75

5.1.1.2   Block-Matching Algorithm Constraint (HC2)

A major portion of the 3D-HEVC ME/DE complexity is related to the TZS algorithm. Although TZS does not compare all reference blocks in a given SW, the computational effort related to TZS remains large in the context of real-time 3D-HEVC encoding. As previously explained, TZS is composed of four main steps: (i) Prediction; (ii) First Search; (iii) Raster; (iv) Refinement. This way, in this analysis some of the TZS steps were disabled, and the different combinations of steps were evaluated. Prediction is the first TZS step, and it employs five different ways to select the SW position in the reference frame: one based on the collocated block (a.k.a., Zero Predictor) and four predictors based on neighboring blocks. By applying predictors based on neighboring blocks, TZS can move the SW to a distant region from the collocated block position and, consequently, increase the memory access overhead. This way, only the Zero Predictor (which does not move the SW) was maintained in TZS for all evaluations.

Additionally, three different levels of TZS simplifications were tested, called TZS1, TZS2, and TZS3, as defined in Table 5.2. The table shows which steps are enabled in each test case. Based on the results of the experiments, the use of TZS without Prediction (except Zero Predictor) and Raster steps was adopted, since this approach presented better trade-off among complexity reduction, BD-Rate increase, and hardware resource usage. TZS3 simplification presents very aggressive image degradation, increasing the BD-Rate in about 40%. TZS1 and TZS2 simplifications present similar BD-Rate results, from 10.14 to 12.09% for video total, and similar total encoding time reduction from 37.74 to 41.10%, respectively. However, by removing the Raster step (TZS2), hardware resource usage, processing, and memory requirements are saved since a considerable number of candidate blocks are avoided. It is important to notice that the Raster step is responsible for testing all candidate blocks inside the search window considering a subsampling parameter. Table 5.3 presents the results in terms of BD-Rate when the constrained TZS is applied over the block-size constraint (HC1). The complete BD-Rate results, according to the 3D-video sequences, can be observed in Appendix C.

Table 5.2 TZS simplifications according to the enabling/disabling steps

TZS steps	BMA constraint		
	TZS1	TZS2	TZS3
Zero predictor only	X	X	X
First search	X	X	X
Raster	X		
Refinement	X	X	

Table 5.3 BD-rate increase according to the block-matching algorithm constraint (HC2)

BMA constraint (HC2)	Type of pictures	Average BD-rate increase (%)		
		1024×768	1920×1088	All sequences
TZS1 + HC1	Video total	4.54	13.50	10.14
	Synth	4.75	12.81	9.79
TZS2 + HC1	Video total	4.85	16.44	12.09
	Synth	5.05	15.57	11.63
TZS3 + HC1	Video total	12.28	55.92	39.55
	Synth	11.11	53.56	37.64

5.1.1.3 Vertical-Disparity Constraint (HC3)

Although 3D-HEVC does not impose any limitation regarding the view arrangement (i.e., the position of the cameras), its coding tools were designed to be more efficient when the views are aligned in 1D linear and coplanar arrangements (Tech et al. 2016). Furthermore, the camera arrangement used to record the videos follows a horizontal displacement (Müller et al. 2014; Tech et al. 2016). Nevertheless, 3D-HTM defines TZS as the BMA applied to the DE prediction and a vertical disparity range equals to 56 by default. In other words, 3D-HTM performs a 2D search in a scenario where only 1D displacement is expected, demanding unnecessary computation by using an inappropriate algorithm for DE prediction.

In summary, given the horizontal-only displacement between cameras, such range is unnecessary for most cases. Therefore, an evaluation considering the horizontal-only disparity was tested, aiming at reducing the memory traffic and its related issues. The strategy increases the accumulated BD-Rate to 13.22 and 14.65% for synthesized views and video total, respectively, when applied over the block-matching algorithm constraint (HC2), as presented in Table 5.4. Based on these results, the use of horizontal disparity was adopted, since it presented a small impact in the accumulated BD-Rate.

5.1.1.4 Additional ME/DE Constraints (HC4)

3D-HTM defines SAD as similarity criterion for the Integer ME/DE, while the Sum of Absolute Transformed Differences (SATD), based on the Hadamard Transform, is applied for the Fractional ME/DE (not used in depth maps), increasing the computational effort. Also, 3D-HTM supports bi-directional prediction and several reference frames for each direction. To reduce the ME/DE computational effort, memory and hardware requirements, three additional modifications were performed: SAD was used for Fractional ME/DE instead of SATD, the support to bi-directional prediction was disabled, and the search was limited to one reference frame per direction (both in ME and DE). The accumulated results show a BD-Rate increase of up to 24.16% when these modifications are applied

Table 5.4 BD-rate increase according to the vertical-disparity constraint (HC3)

Sequence	Average BD-rate increase (%) according to the HC3 (vertical-disparity constraint + HC2)	
	Video total	Synthesized
Balloons	5.76	5.74
Kendo	9.08	8.02
Newspaper_CC	6.20	5.47
GT_Fly	26.84	26.10
Poznan_Hall2	18.81	14.45
Poznan_Street	17.11	13.20
Undo_Dancer	8.86	8.06
Shark	24.58	24.72
1024×768	7.01	6.41
1920×1088	19.24	17.31
All sequences	14.65	13.22

over the vertical-disparity constraint (HC3), as presented in Table 5.5. When these additional ME/DE constraints were implemented in the 3D-HTM reference software, the total encoding time was reduced by 59%. Based on that, the additional ME/DE constraints were adopted due to the great reduction in the 3D-HEVC ME/DE complexity.

Table 5.5 BD-Rate increase according to the additional ME/DE constraints (HC4)

Sequence	Average BD-Rate increase (%) according to the HC4 (additional ME/DE constraints + HC3)	
	Video total	Synthesized
Balloons	7.21	8.38
Kendo	21.51	19.86
Newspaper_CC	6.90	7.14
GT_Fly	42.96	42.23
Poznan_Hall2	22.89	20.08
Poznan_Street	18.99	17.53
Undo_Dancer	17.31	16.10
Shark	55.51	54.47
1024×768	11.87	11.79
1920×1088	31.53	30.08
All sequences	24.16	23.22

5.1.1.5 ME/DE Constraint Analyses Summary

Table 5.6 summarizes the BD-Rate increase of all ME/DE hardware-oriented constraints. One may notice that by using all constraints (HC4), BD-rate is increased by 23.22 and 24.16% for synthesized views and video total, respectively. The simplifications were implemented in the 3D-HTM reference software, and the encoding computational effort was reduced by 59%. Therefore, the modified 3D-HEVC with the applied ME/DE constraints was adopted as the Baseline Encoder for the developed Energy-Aware 3D-HEVC ME/DE System presented in the next section. The complete BD-Rate results, according to the 3D-video sequences, can be observed in Appendix C.

The accumulated coding-efficiency losses are not negligible, but the constraints are necessary when real-time encoders for mobile devices are considered. The need for simplifications becomes clear when we observe that related works focusing on HEVC (less complex than 3D-HEVC) also employ similar constraints, such as a limited number of PU sizes (Afonso et al. 2013; Afonso et al. 2016), simplified BMAs (Doan et al. 2017; He et al. 2015; Pastuszak et al. 2016b), SAD as the unique similarity criterion (Pastuszak et al. 2016b), and a limited number of reference frames (Pastuszak et al. 2016b). Unfortunately, most of these works do not present detailed results regarding the losses in terms of compression efficiency due to the adopted simplifications. Other works present comparisons with configurations that do not consider all encoding tools available in the reference software by default so that the full impact of the simplifications cannot be estimated. In the developed hardware design, the ME/DE system was developed after the base configuration is carefully defined and characterized.

Table 5.6 BD-Rate increase according to the ME/DE hardware-oriented Constraint (HC)

HC	Constraints	Type of pictures	Average BD-rate increase (%)		
			1024×768	1920×1088	All sequences
1	Block sizes (16×16 and 32×32)	Video total	3.81	8.92	7.00
		Synth	4.19	8.44	6.85
2	BMA (TZS without *Predictors* and *Raster*) + HC1	Video total	4.85	16.44	12.09
		Synth	5.05	15.57	11.63
3	Horizontal disparity only + HC2	Video total	7.01	19.24	14.65
		Synth	6.41	17.31	13.22
4	Additional ME/DE constraints + HC3	Video total	11.87	31.53	24.16
		Synth	11.79	30.08	23.22

5.1.2 Developed 3D-HEVC ME/DE System

In order to make clear the understanding of the system development, this section is divided into two subsections. The first one presents an overview of the developed energy-aware ME/DE system, whereas the second describes the design of the dedicated processing units. After, Sect. 5.1.3 shows the developed hardware-oriented BMA and scheduling, and Sect. 5.1.4 presents the developed on-chip memory design and management.

5.1.2.1 Energy-Aware ME/DE System Overview

Figure 5.1 shows the diagram of the developed energy-aware 3D-HEVC Motion and Disparity Estimation architecture. The architecture is composed of three main hardware units, one responsible for processing the Base View (BV) and two for the Dependent Views (DVs; due to space restrictions only one instance is depicted in Fig. 5.1). The Base View, or Independent View, is the first one to be encoded and it does not depend on other views, allowing compatibility with 2D HEVC-based systems. The remaining ones are Dependent Views, i.e., those whose encoding process depends on information from other views. The Base-View unit has one ME processing unit per channel (texture and depth), whereas each Dependent-View unit has one ME and one DE processing unit per channel. Each processing unit has a private on-chip SRAM memory. Every processing unit and on-chip memory features specific optimizations and observe distinct requirements according to the data characteristics, processing requirements, and available information from previously processed data. Figure 5.1 summarizes the techniques applied to each unit, which are detailed in Sects. 5.1.3 and 5.1.4.

To guarantee HEVC backward compatibility, it is mandatory for the Base View to process the texture channel before the depth channel. Observing this restriction and aiming at efficiently exploiting the inter-channel correlation, the Texture unit provides valuable information related to the ME result to the Depth-Map unit that, under specific situations, may completely skip the search process according to the proposed Texture-Based Motion Vector Inheritance (TMVI) algorithm (Sect. 5.1.3.3). To further exploit the inter-channel correlation, the proposed system employs the Flexible Coding Order for the Dependent Views. This allows the Depth-Based Dynamic Search Window Resizing (DSWR) to adapt the SW with proper knowledge of the scene as both disparity and motion displacements are related to depth (Sect. 5.1.3.4). The red arrows in Fig. 5.1 represent auxiliary information flow used by the proposed algorithms.

The proposed system is implemented in a 4-stage macro-pipeline at the frame level, as depicted in Fig. 5.2. The first stage processes texture for the Base View and the second processes depth for the Base View. In turn, the third and fourth stages process (in parallel) ME and DE of all Dependent Views for depth maps and texture, respectively. As the number of cycles spent in each stage varies according to data characteristics, once a unit has concluded its task, clock gating is applied for energy reduction.

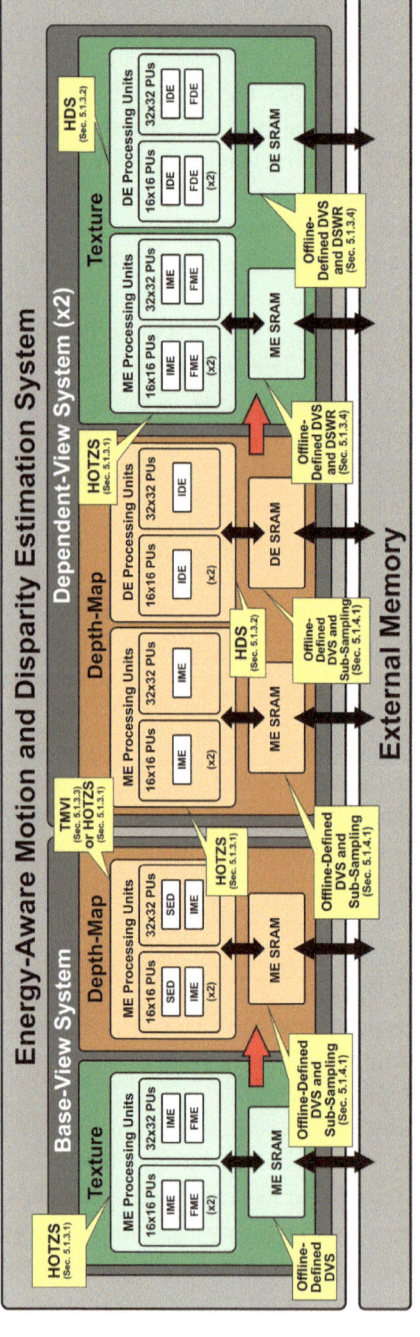

Fig. 5.1 Energy-aware ME/DE system (Adapted from Afonso et al. (2019a))

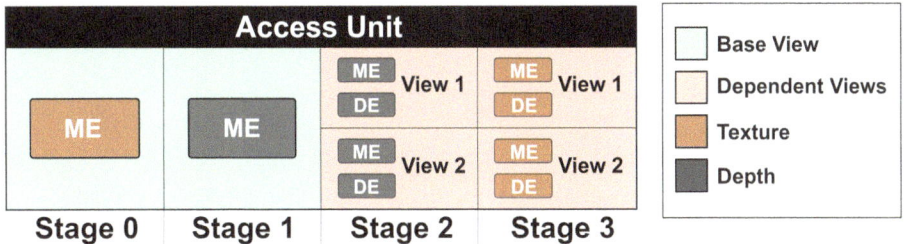

Fig. 5.2 ME/DE system macro-pipeline (Adapted from Afonso et al. (2019a))

5.1.2.2 Processing Units Design

The dedicated processing units were developed according to average-case design strategy observing the 3D-HEVC reference software behavior. Each processing unit was defined to process 16 lines in parallel, where each line is composed of 16 or 32 samples, depending on the block size (Sect. 5.1.1.1). Whenever 16×16 blocks are processed, part of the logic circuit is clock-gated. Based on a statistical analysis of the number of candidate blocks during the BMA process, 240 candidates was defined as the target performance. Such performance was calculated considering the adoption of the previously described constraints.

ME/DE processing units present a similar behavior. They both divide each 64×64 CTU into four 32×32 CUs, sequentially processing each one. The CU is evaluated as one 32×32 block and as four 16×16 blocks. At the beginning of the process, the 32×32 and both upper 16×16 blocks begin to be processed in parallel. Immediately after, the remaining ones start to be processed in parallel, as detailed in Fig. 5.3.

Figure 5.4 presents the architecture for the operative part of a 16×16 ME processing unit (a similar structure is used for DE and 32×32 units) used in the coding process of texture channel for the Base Views. Figure 5.5 details the basic hardware structures that compose the processing unit: (a) SAD tree; (b) SAD accumulator; (c) Type-A interpolation filter; (d) SAD comparator; and (e) Type-B interpolation filter. The interpolation filters are used in the Fractional Motion Estimation (FME) processing. This step is applied

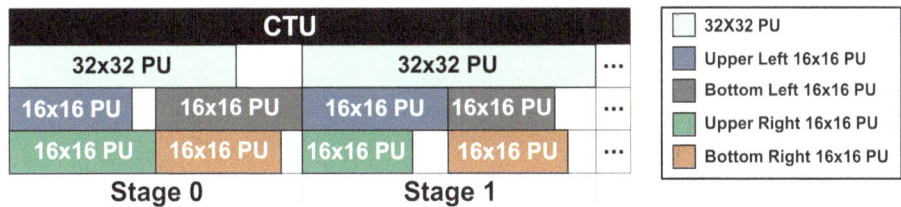

Fig. 5.3 ME/DE Block-encoding process macro-pipeline (Adapted from Afonso et al. (2019a))

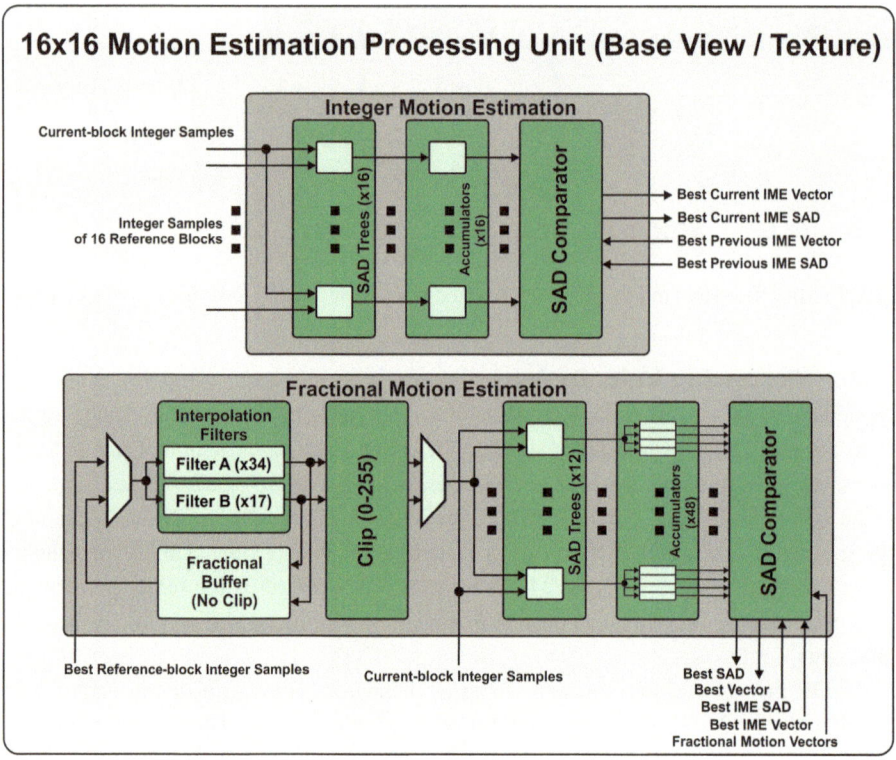

Fig. 5.4 16×16 ME processing unit used in the texture processing for the base view (Adapted from Afonso et al. (2019a))

to the best candidate found by the IME in order to improve the coding efficiency. By default, FME is not applied to the depth map.

5.1.3   Hardware-Oriented BMA and Scheduling

This section is divided into four subsections. The first one presents the Hardware-Oriented TZS static scheduling (HOTZS). The second shows the developed Horizontal Disparity Search (HDS) algorithm. The third describes the developed Texture-Based Motion Vector Inheritance (TMVI) algorithm. Finally, the fourth subsection presents the Depth-Based Dynamic Search Window Resizing (DSWR) algorithm.

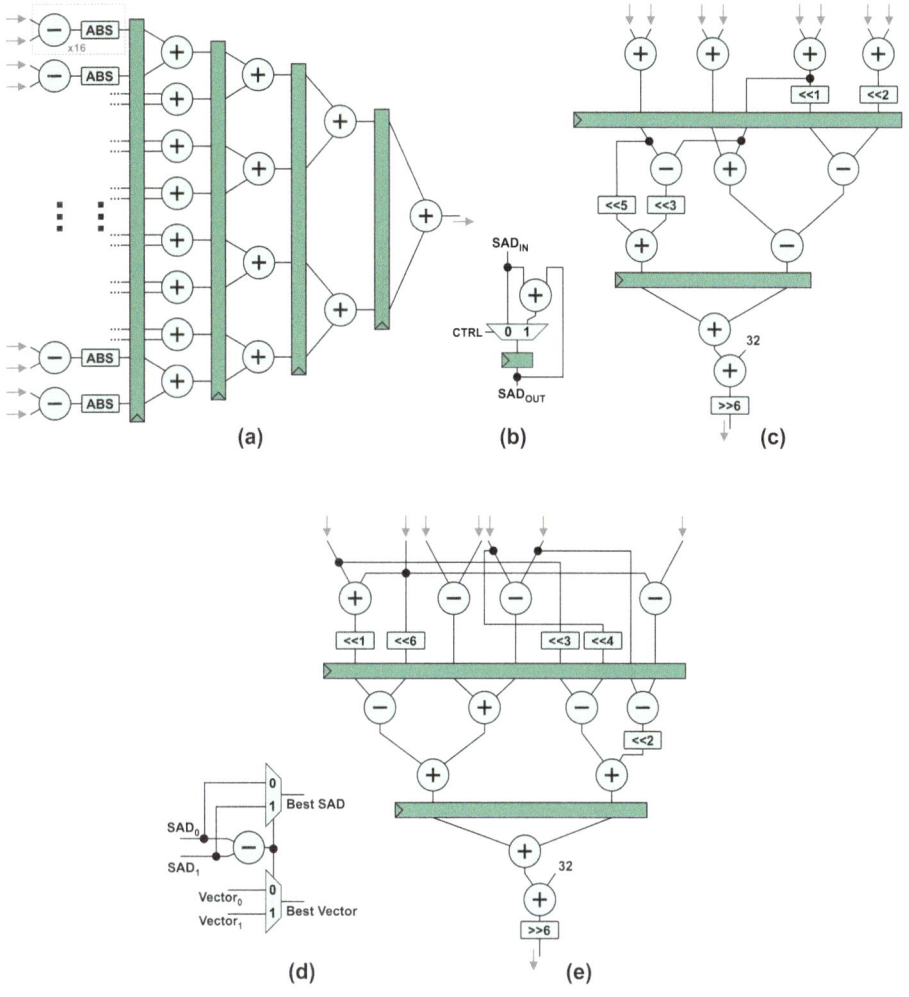

Fig. 5.5 Basic hardware structures: **a** SAD Tree of 16×16 processing units; **b** SAD accumulator; **c** Type-A interpolation filter; **d** SAD comparator of two blocks; **e** Type-B interpolation filter (Adapted from Afonso et al. (2019a))

5.1.3.1 Hardware-Oriented TZS Static Schedule (HOTZS)

As described in Sect. 5.1.1.2, the TZS algorithm was modified by removing two main steps. By removing the Prediction step, search window displacement is avoided, and data reuse is improved. Computational effort is reduced by removing the Raster step. On top of that, a static TZ search scheduling (HOTZS) to provide a more regular memory access pattern and improve the Processing Units usage was also defined. As it can be observed in Fig. 5.6a, the TZS is composed of multiple iterations composed of 1, 4, 8 or 16 candidate

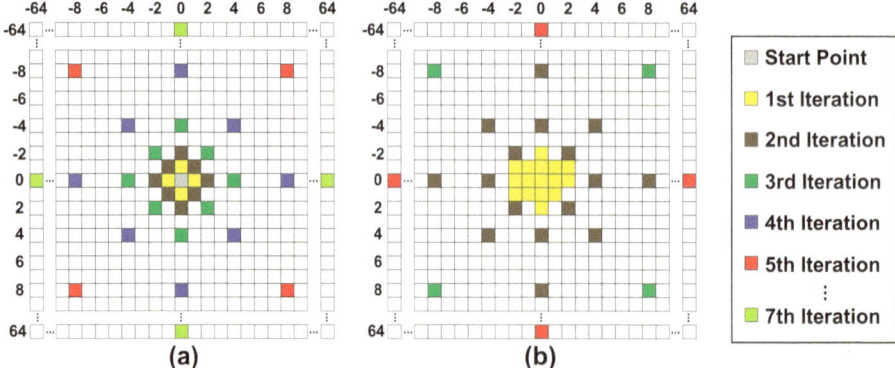

Fig. 5.6 Processing of candidates according to each iteration: **a** TZS; **b** HOTZS (Adapted from Afonso et al. (2019a))

blocks each. To use the 16-line parallelism, HOTZS rearranges the processing schedule to process 16 candidates per iteration, as shown in Fig. 5.6b.

5.1.3.2 Horizontal Disparity Search (HDS)

Since 3D-HEVC is meant to encode horizontally displaced video streams, the diamond-shaped search pattern employed by TZS leads to two drawbacks during the DE process: (i) unnecessary computation and increased search window, resulting from vertical block matchings; (ii) local minima trapping when TZS converges in the first iterations, leading to inaccurate disparity field estimation.

Therefore, the Horizontal Disparity Search (HDS) algorithm, composed of three steps, was proposed to exploit the horizontally displaced video characteristic. The First Step consists of a horizontal search from the start point comparing the current block with candidate blocks eight samples apart from each other (i.e., a subsampled horizontal-only raster search), represented in the upper part of Fig. 5.7 by the colored boxes in the First Step. Similarly, the Second Step performs the horizontal search, but the distance between candidates is two samples. The Second Step is centered at the best block matching of the First Step (red box in the example of Fig. 5.7). Finally, the Third Step performs a bidirectional search around the best matching obtained from the Second Step. Although there is no real vertical disparity, a 1-sample vertical search refinement is employed to assure good matching under slight cameras vibrations or captured noise. The HDS allows reducing the 192×192-sample memory for DE to a 192×66-sample memory. Additionally, the number of block matchings during HDS is constant, simplifying memory-access pattern and processing-unit design.

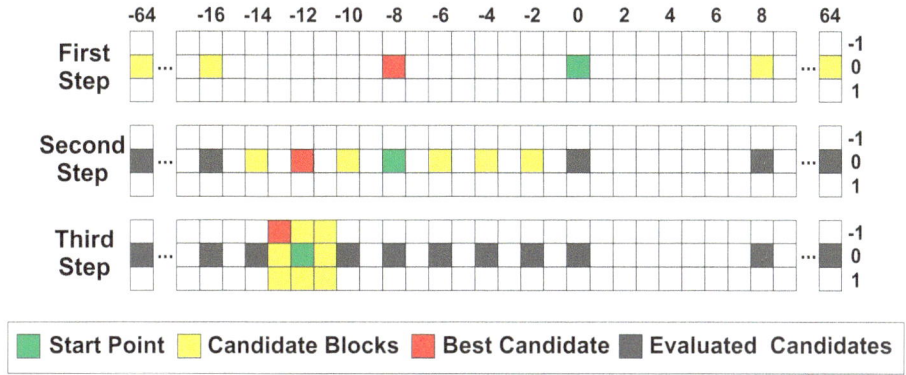

Fig. 5.7   Horizontal disparity search algorithm (Adapted from Afonso et al. (2019a))

5.1.3.3   Texture-Based Motion Vector Inheritance (TMVI)

Texture-Based Motion Vector Inheritance (TMVI) was implemented aiming at exploring the correlation between texture and depth map for the Base View. Although there is a correlation between the texture and the depth channels, texture information does not exist in depth maps. This way, different motion vectors lead to good matchings for homogeneous regions of a depth map (regions generally inside the objects), which includes the motion vector selected during the texture encoding of the collocated block.

The TMVI calculates the Simplified Edge Detector (SED) metric (Sanchez et al. 2014a) to define if the depth block under processing is a sharp-edge or homogeneous region. In the case of sharp-edge detection, the HOTZS is normally applied to the depth map ME. Otherwise, for homogeneous regions, the motion vector used for the collocated block in the texture channel is inherited by the depth block. Thus, a complete search is skipped (leading to block matching complexity reduction) and the memory associated with the depth ME at the Base View may be scaled to idle mode.

5.1.3.4   Depth-Based Dynamic Search Window Resizing (DSWR)

DSWR is implemented to dynamically reduce the search window by exploiting depth information in the texture ME and DE processes. Typical implementations employ a constant search window along the whole ME/DE process, disregarding the block/object depth within the scene, as shown in Fig. 5.8a. During the ME process, DSWR exploits the fact that the displacement (motion vectors) of objects/blocks moving at the same speed is inversely proportional to the distance between object and camera (depth), i.e., the closer the object, the larger the motion vector (for objects moving at same speed) and, thus, a larger SW is necessary. The same applies to DE estimation, i.e., closer objects present larger disparity vectors.

Therefore, DSWR scales the active SW (shaded area in Fig. 5.8) according to a perspective projection, as depicted in Fig. 5.8b. Additionally, DSWR considers the disparity/

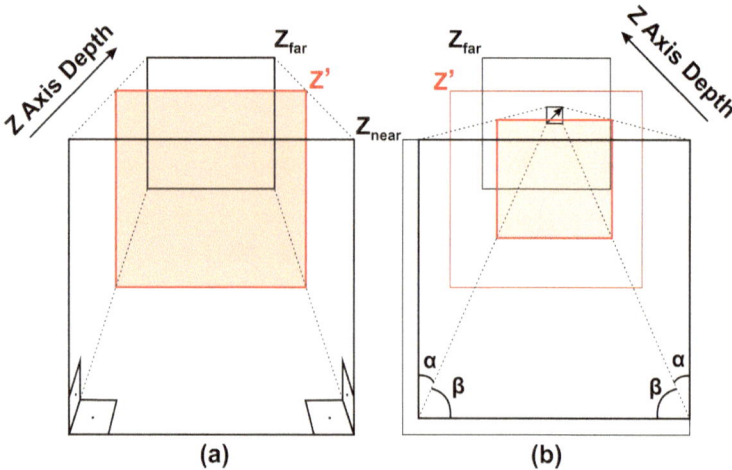

Fig. 5.8 Search window size: **a** Fixed search window size; **b** depth-based dynamic search window resizing (Afonso et al. 2019a)

motion vectors selected during the depth-map ME/DE to displace the active SW by applying a projection distortion. This behavior is based on the observation that collocated blocks of the same view (texture and depth) represent the same scene projection and, consequently, tend to present similar motion and disparity vectors. In Fig. 5.8, the Z_{near} and Z_{far} correspond to the depth-sample values of the closest and farthest objects in the scene respectively. Z' corresponds to the current block depth value to be considered for the DSWR resizing.

Figure 5.9 presents the algorithm used to define the SW resizing simplified to one dimension (horizontal in this case). First, the horizontal component of the vector (MV_x) of the collocated block is obtained (line 2). Afterward, the average depth of the collocated depth-map block is stored in the S variable (line 3). Then, an offset is added to the module of MV_x and stored in the absMV (line 4). After, the vector orientation is extracted (line 5). The search range right side (srchRngPosSide) is defined using Z_{near} and S (lines 6–7). Similarly, the search range left side (srchRngNegSide) is defined (lines 8–12). Finally, the Search Range is placed according to the vector orientation (lines 13–19) and mapped to an SW rounded to multiple of 16 samples (lines 20–21) to fit the memory sector organization.

1. //Define the X dimension of texture search range
2. MV$_x$ ← getCorrelatedDepthVector(); // get correlated vector in depth map
3. S ← getCorrelatedDepthAverage(); //get correlated depth-map block average
4. absMV ← abs(MV$_x$) + OFFSET; //insert an offset in the obtained vector
5. signal ← (MV$_x$ ≥ 0) ? 1 : -1; //save the obtained vector signal
6. a ← $\frac{(64 - absMV)}{z_{near}}$; //define the a constant
7. srchRngPosSide ← a × (S − Z$_{near}$) + 64; //define size of positive side of SR
8. **If**(MV$_x$ ≠ 0)**Then** //define the size of negative side of SR
9.   srchRngNegSide ← srchRngPosSide–absMV;
10. **Else**
11.   srchRngNegSide ← srchRngPosSide;
12. **End If**
13. **If**(signal = 1)**Then** //define the new search range
14.   SR$_{RIGHT}$ ← srchRngPosSide;
15.   SR$_{LEFT}$ ← srchRngNegSide × (-1);
16. **Else**
17.   SR$_{RIGHT}$ ← srchRngNegSide;
18.   SR$_{LEFT}$ ← srchRngPosSide × (-1);
19. **End If**
20. SW$_{RIGHT}$ ← roundSrchWin(SR$_{RIGHT}$); //round the new search window
21. SW$_{LEFT}$ ← roundSrchWin(SR$_{LEFT}$);

Fig. 5.9 DSWR algorithm pseudo-code (Afonso et al. 2019a)

5.1.4   On-Chip Memory Design and Management

This section is divided into two main subsections. The first subsection presents the memory organization and sizing, and the second subsection shows the techniques proposed for memory management.

5.1.4.1   Memory Organization and Sizing

The proposed system employs four different types of on-chip memories. However, both ME and DE memories follow the same memory organization presented in Fig. 5.10, independently of the channel (texture or depth). The ME on-chip memories cover an SW of 192 × 192 samples and are composed of 144 sectors, organized in 12 columns and 12 rows. Each sector stores 16 128-bit words, i.e., each word is composed of 16 samples of 8 bits for texture. As the main benefit of our HDS algorithm (Sect. 5.1.3.2), the DE memories only need to store 192 × 66 samples within 72 sectors, organized in 12 columns and six rows (upper and bottom sectors distributed in the six rows store only one additional word). By applying the Depth Sub-Sampling technique presented in the following, which reduces the sample representation to 4 bits, the memories for depth maps comprise 64-bit lines.

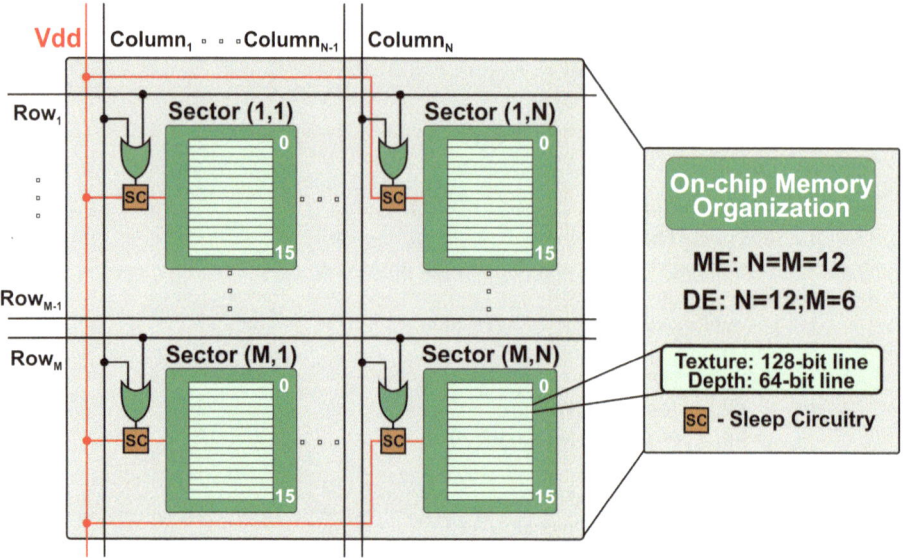

Fig. 5.10 Organization of on-chip memory units (Adapted from Afonso et al. (2019a))

In the MVD format, depth map pixels are represented using one channel composed of 8-bit samples. Additionally, depth map content is dominated by smooth regions (within objects and background) and very sharp edges (object borders), i.e., there are no complex-textured regions. Observing these characteristics, a bit-depth subsampling is implemented to reduce the size of on-chip memories for depth maps by 50% (subsampling from 8 bits down to 4 bits) with small coding efficiency drawback. For that, four sub-sampling patterns were evaluated. The evaluated patterns remove the (i) four more significant bits; (ii) four less significant bits; (iii) odd bits; and (iv) even bits, as presented in Table 5.7. As the elimination of the four less significant bits demonstrated the lowest drawback in coding efficiency (0.59% in BD-Rate), it was adopted for the design of depth maps memories. This evaluation used the experimental setup shown in Sect. 5.1.1.

Table 5.8 presents the total memory size for each memory type. Note that DE memories are smaller than ME memories, and depth-map memories are smaller than texture memories due to the proposed techniques.

5.1.4.2   Run-Time Adaptive Memory Management

The following paragraphs present the power model and the proposed run-time adaptive memory management.

Sector-level dynamic voltage scaling (DVS) is applied to reduce leakage energy in idle sectors. The power state management is performed using a power-state matrix signaled by $Column_x$ and Row_y in Fig. 5.10. The Sleep Circuitry (SC in Fig. 5.10) is responsible for

Table 5.7 BD-Rate increase according to the depth maps sub-sampling

Sub-sampling strategy	Type of pictures	Average BD-rate increase (%)		
		1024×768	1920×1088	All sequences
SubSamp = *Samp & 0 × 0F*	Video total	0.28	2.52	1.68
	Synth	1.16	3.87	2.85
SubSamp = *Samp & 0 × F0*	Video total	−0.06	0.50	0.29
	Synth	0.21	0.81	0.59
SubSamp = *Samp & 0 × 55*	Video total	−0.18	0.94	0.52
	Synth	0.31	1.60	1.12
SubSamp = *Samp & 0 × AA*	Video total	−0.06	1.33	0.81
	Synth	0.34	1.87	1.30

Table 5.8 Memory sizing

Memory	#samples	#bits per sample	#bytes per instance	#sectors (rows × columns)	Instances	Total size
ME—Texture	192×192	8	36.86 kB	144 (12 × 12)	3	110.58 kB
ME—Depth Map	192×192	4	18.4 3 kB	144 (12 × 12)	3	55.29 kB
DE—Texture	192×66	8	12.67 kB	72 (6 × 12)	2	25.34 kB
DE—Depth Map	192×66	4	6.34 kB	72 (6 × 12)	2	12.68 kB
Total	–	–	**74.3 kB**	–	–	**203.89 kB**

actuating on the sector voltage (V_{dd}) switching between three power states: PS_{on} ($V_{dd} = 1V$), PS_{idle} ($V_{dd} = 0.6V$), and PS_{off} ($V_{dd} = 0V$), where PS_{on} and PS_{idle} are data retentive. The nvsim (NVSIM 2024) simulator along with the 45nm CMOS technology power models (Wang et al. 2015) was used to estimate the energy consumption for leakage, reading, writing, and power state transition, as summarized in Table 5.9. The wakeup latency was considered negligible for this analysis.

As previously mentioned, Depth-Based Dynamic Search Window Resizing (DSWR—Sect. 5.1.3.4) and Texture-Based Motion Vector Inheritance (TMVI—Sect. 5.1.3.3) are proposed to allow dynamic power management through DVS of idle memory regions taking advantage of the inter-channel correlations. DSWR dynamically reduces the search-window during texture ME/DE processing based on the depth ME/DE encoding decisions for the Dependent Views, disabling texture-memory sectors based on the depth-block motion vector and sample average. TMVI prioritizes the use of ME/DE computational

Table 5.9 Power states energy

Power state	Vdd (V)	Energy (J)					
		Per bit			Per transition		
		Leak	Write	Read	PS_{on}	PS_{idle}	PS_{off}
PS_{on}	1.0	8.467f	0.082p	0.127p	–	0	0
PS_{idle}	0.6	4.472f	–	–	0.062f	–	
PS_{off}	0	0	–	–	0.113f	0.038f	–

effort over the sharp-edge regions of depth maps. This way, for homogeneous regions, TMVI inherits the motion vector used for the collocated block in texture channel, disabling the depth-memory sectors.

5.1.5   Results

In order to make the discussion of results clearer, this section was divided into four sub-sections. Firstly, the run-time adaptive memory hierarchy results are presented, followed by the processing-unit results, system results, and the comparison with related works.

5.1.5.1   Adaptive Memory Results

To perform a fair energy consumption analysis for the on-chip memories, a memory simu-lator capable of emulating their behavior was developed (in C#; VITECH 2024), allowing the collection of the following numbers for each memory sector: cycles on, cycles sleep and toggles. Additionally, the total number of readings and writings were collected for each memory line. The simulator input is a trace file indicating all memory requests extracted using the 3D-HTM reference software during the ME/DE processing. It is worth saying that two traces were generated: (i) using 3D-HTM considering only hardware-oriented simplifications (Sect. 5.1.1); (ii) considering all techniques proposed in this book (Sect. 5.1.2). The use of two different traces is needed since the proposed algorithms change the BMA behavior leading to different memory requests. The energy consumption results provided by the simulator were estimated considering the power model presented in Sect. 5.1.4.2.

Three memory scenarios were considered for evaluation and were applied on top of the B-Encoder configuration. The Base Memory Hierarchy (BMH) consists in the proposed hierarchy without memory size reduction techniques, adaptive management algorithms or data reuse; the Reduced-Size Hierarchy (RSH) features the proposed memory hier-archy with size reduction but without adaptive algorithms; and the Run-Time Adaptive Hierarchy (RAH) comprises the proposed memory hierarchy with run-time adaptive man-agement. For any of these memory scenarios, the impact in BD-Rate was obtained with the same experimental environment presented in Sect. 5.1.1. It is important to notice that BMH presents the same BD-Rate increase of the B-Encoder (see Sect. 5.1.1.5), RSH increases the BD-Rate by up to 0.59% when compared to the B-Encoder (see Table 5.7), and RAH increases the BD-Rate by 1.54 and 1.47% for synthesized views and video total, respectively (when compared to the B-Encoder). The complete BD-Rate results, according to the 3D-video sequences, can be observed in Appendix C.

Figure 5.11a presents a bar chart of the power dissipation for each of the 10 ME/ DE memories in the three mentioned scenarios (DV and BV stand for Dependent and Base Views, respectively). As expected, the BMH solution presented the highest power dissipation (2.53W). Furthermore, one may notice that the memory size reduction (RSH solution) leads to a power dissipation of 1.86W, which means a decrease of 26.5%. The RAH solution led to a power dissipation of 1.10W, i.e., a reduction of 56.5% when compared to BMH. Additionally, Fig. 5.11b presents the energy consumption distribution of each experiment considering all AUs. Note that, in addition to energy reduction, the Run-Time Adaptive Hierarchy leads to a more concentrated distribution, demonstrating the low energy consumption along different video sequences and QPs.

It is worth mentioning that the contributions regarding on-chip memory hierarchy and management consider SRAM memories. However, these contributions focused on the architectural and control levels, so the significant gains achieved by the RAH over RSH and BMH memory hierarchies are independent of the technology. Thus, the proposed solution can be applied to on-chip memory technologies other than SRAM.

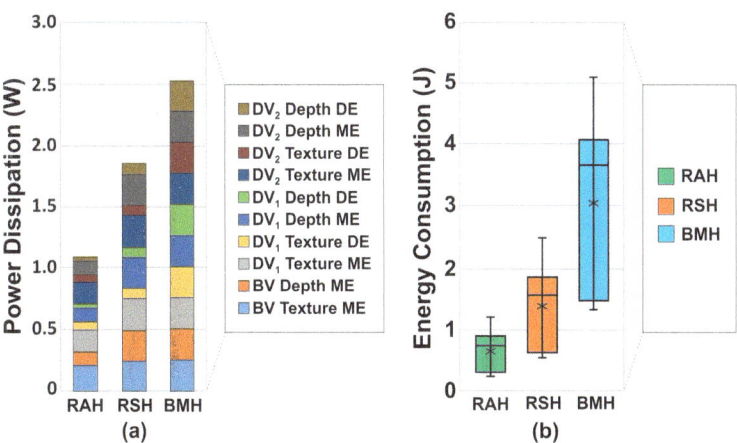

Fig. 5.11 On-chip memory results: **a** Power dissipation breakdown and **b** energy consumption boxplot (Adapted from Afonso et al. (2019a))

Table 5.10 Hardware design synthesis results

Architecture	Area (μm^2)	#Gates ($\times 10^6$)	Power (W)			Instances
			Leak	Dyn.	Leak. + Dyn.	
ME—Texture	8,542,735	6.817	0.147	0.984	1.131	3
ME—Depth Map	2,797,576	2.232	0.047	0.422	0.469	3
DE—Texture	6,388,479	5.098	0.123	0.509	0.631	2
DE—Depth Map	2,220,003	1.772	0.041	0.152	0.193	2
SubTotal	19,948,793	15.919	0.358	2.067	2.424	–
*Total	51,237,897	40.888	0.908	5.539	6.447	–

*The total results consider all instances of ME/DE architectures

5.1.5.2 Processing Units Synthesis Results

The developed hardware was synthesized targeting an ASIC technology according to the experimental setup presented in Appendix B. Table 5.10 presents the synthesis results obtained for each design (ME/DE and texture/depth). It is important to highlight that the results for each ME/DE architecture consider only one instance of the processing units. In other words, each result is related to the ME or DE processing of one channel, texture or depth, and the number of instances needed in the system varies according to the encoding tool and the channel, as previously discussed in Sect. 5.1.2.1. As expected, both texture architectures consumed a more on-chip area compared to depth ones due to the internal bit depth. The full hardware design that considers all instances of the ME/DE architectures has 51.2mm^2 of on-chip area (40.9M Gates, considering the NAND-2 gate size) while presenting a total dissipation of 6.45W@100MHz.

5.1.5.3 System Results

Figure 5.12 presents a comparison regarding energy consumption per AU between the three scenarios considered in the developed architecture: (i) BMH, (ii) RSH, and (iii) RAH. Note that the BMH implementation presented a mean energy consumption of 0.37J per AU, whereas RSH consumes 0.19J (reduction of 48.6%). Finally, the developed Energy-Aware Motion and Disparity Estimation system, considering all proposed techniques, presents average consumption of only 0.107J per AU, thus reducing the total energy consumption by 71.1% in comparison to the BMH implementation, with a BD-Rate increase of only 1.54%.

Current commercial mobile devices include on-chip video codecs in addition to other modules responsible for specific functionalities or general purpose. However, manufacturers do not release detailed results of their hardware accelerators. Furthermore, works that present a power analysis for the different components implemented inside a device are rarely found in the literature, mainly considering current devices and current video-coding standards. These facts avoid that a comprehensive analysis regarding power savings by

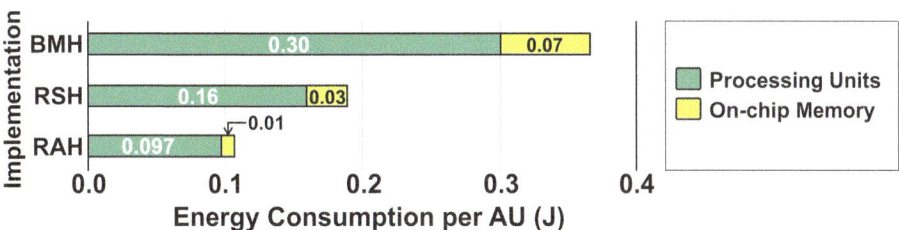

Fig. 5.12 Comparison among base memory hierarchy, reduced-size hierarchy, and run-time adaptive hierarchy energy consumption per AU (Adapted from Afonso et al. (2019a))

the proposed methodology in comparison with the overall power dissipated by a mobile device is made.

The work (Carroll et al. 2013) presents a power analysis of the Samsung Galaxy S III smartphone (launched in 2012; Samsung 2024), in which the power consumption is measured by instrumentation at circuit level for the major components. Also, this work analyzes power consumption while decoding two 1280×720 (HD 720p) H.264 videos. For each video, the authors measured the consumption of the hardware video decoder and the software decoder. For a high-quality HD 720p video, more than 1W is saved by using the dedicated video compression hardware, reducing about 45% of the overall power dissipation. It is important to notice that a recording process (considering the video encoding instead of the decoding) demands higher power dissipation and up to 98% of the total energy consumption is related to the ME/DE steps (Zatt et al. 2011b) considering off-chip memory access, on-chip memory and computation under the MVC standard. In the context of 3D-HEVC, it is also expected that the major part of the energy consumption is due to the ME/DE steps. Therefore, as the main proposed strategy reduces the energy consumption by 71.1% compared to BMH (related to the on-chip memory and computation), it is possible to infer that the power dissipation saved is relevant when current mobile devices are considered.

5.1.6   Main Related Works and Comparisons

As mentioned before, there was no related works proposing dedicated systems for 3D-HEVC ME/DE found in the literature. Even though a direct comparison is not possible, this section compares the proposed system with systems designed for previous standards. The on-chip memory solution proposed in (Zatt et al. 2011b) can achieve up to 65% of energy consumption in comparison to Level-C. The Reduced-Size Hierarchy solution presents an average energy-consumption reduction of 79% when compared to Base Memory Hierarchy (which basically employs the Level-C). The Multi-view Video Encoder proposed in (Ding et al. 2010) dissipates 366mW@166MHz when processing two-view

1920×1080 3D videos. Although presenting less power dissipation when compared to the developed work, note that the solution proposed in (Ding et al. 2010) was built for a previous and less complex standard (MVC) where only two views (without depth maps) are considered, while the system developed in this book allows the processing of three texture frames and their respective depth maps. In other words, the system developed in this book enables the viewing of multiple views at the decoder side after view synthesis. Finally, the solution in (Sampaio et al. 2013) reduces the energy consumption by 77–88% when compared to the Level-C approach (considering the MVC standard), whereas the developed system reaches a reduction of 79% for 3D-HEVC compared to Level-C.

In conclusion, the developed 3D-HEVC Motion and Disparity Estimation system was designed for low-energy consumption featuring a dedicated memory hierarchy capable of run-time adaptation. Through several proposed algorithms on-chip memory reduction and dynamic power management were allowed. Three different memory hierarchy approaches were developed aiming at reducing energy consumption. Experimental results demonstrated an average energy consumption reduction of 79% for the memory when compared to the widely used Level-C scheme. By using the run-time adaptive management combined with the memory-size reduction techniques, the total ME/DE energy consumption is reduced by 71.1%. The developed architecture is the only solution able to process ME/DE for 3D-HEVC supporting up to three HD 1080p views (three texture views and their associated depth maps).

5.2   Low-Power and Coding-Efficient Disparity Estimation Architecture Based on Improved Unidirectional Disparity-Search Algorithm

Few works proposing hardware solutions for the DE step can be found in the literature and none of them was focusing on the 3D-HEVC. The works (Fan et al. 2018) and (Perleberg et al. 2018) successfully proposed hardware solutions for the ME step of the HEVC standard considering modified versions of the Test Zone Search (TZS) algorithm (Li et al. 2014b), which are compatible to the 3D-HEVC DE step. But solutions able to jointly propose hardware-friendly algorithms and efficient hardware architectures are still required to allow real-time processing of all 24 PU sizes defined by the 3D-HEVC. In this book, an intensive effort is done in order to overcome these gaps and a low-power and coding-efficient Disparity Estimation architecture based on a novel algorithm is proposed, as follows.

5.2.1 An Efficient Low-Complexity BMA for the 3D-HEVC DE

As previously explained, the DE step is responsible for reducing the Inter-view redundancy, in a computationally intensive process. To constrain the DE computational effort, the BMA is applied within a pre-defined and reduced search area (SA), which size depends on the size of the CTU and the user-defined Search Range (SR). There are several BMAs in the literature designed focusing on the ME step which are also used in the DE step (Fan et al. 2018; Perleberg et al. 2018). The 3D-HEVC Reference Software, the 3D-HTM (3D-HEVC Reference Software 2024), adopts the TZS to perform both, ME and DE steps. TZS can maintain near-optimal performance whereas reducing $23 \times$ the computational effort when compared to the Full Search (FS) algorithm (Li et al. 2014b). However, once it was designed considering ME characteristics, TZS processes candidates in both horizontal and vertical positions related to the TZS start point. This way, TZS is unable to take advantage of DE specificities because the candidates at vertical positions have a low probability of being selected to represent the disparity. This way, BMAs focusing on the processing of candidates located in horizontal positions are more efficient to be used in the DE step since the 3D-HEVC was developed focusing on views captured with horizontal displacement (Tech et al. 2016; Müller et al. 2014). The diamond-shaped search pattern employed by TZS leads to two drawbacks during DE process: (i) unnecessary computation and increased search window; (ii) potential local minima trapping when the TZS converges in the first iterations. Therefore, this book proposes an efficient low-complexity BMA for the 3D-HEVC DE, the Unidirectional Disparity-Search (UDS) algorithm. The developed algorithm is composed of three steps, and it considers a multi-view approach with a horizontal camera arrangement, as presented in the following. However, it is important to highlight that the algorithm can be easily adapted for a possible vertical camera arrangement.

5.2.1.1 UDS: Unidirectional Disparity Search Algorithm

The first UDS step, called the First Step, consists of a horizontal search from the start point (collocated block in the reference frame), which compares candidate blocks with a block sub-sampling defined by the S_1 parameter. The value of S_1 was chosen based on a wide offline evaluation using the 3D-HTM reference software as will be discussed in the following. Figure 5.13 presents an example with $S_1 = 12$, i.e., a block sub-sampling of 12:1, where the green circle represents the start point, the yellow circles represent the candidate blocks, and the red circle represents the best candidate. The First Step considers all available horizontal candidate blocks inside the search-window, concerning the S_1 block sub-sampling criteria.

Similarly, the Second Step also performs a horizontal search, but with a lower sub-sampling rate, defined by S_2. This sub-sampling factor was also defined based on evaluations with the 3D-HTM. $S_2 = 4$ is used as an example in Fig. 5.13. The Second Step is centered at the best block matching of the First Step, and it is performed until

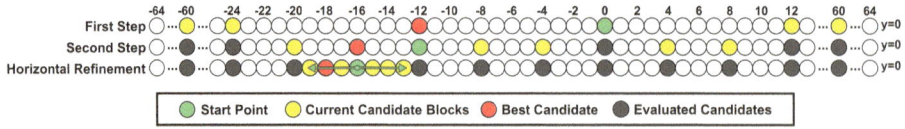

Fig. 5.13 Steps of the proposed UDS algorithm (Adapted from Afonso et al. (2018))

reaching the search window limit or the limits given by the two neighboring candidate blocks evaluated in the previous step. The gray circles represent the candidate blocks evaluated along previous steps of the UDS algorithm. Finally, the third step consists of a Horizontal Refinement centered at the best block matching of the Second Step, and it considers all candidate blocks between the two neighboring evaluated candidates from the Second Step, as illustrated in Fig. 5.13.

The UDS algorithm evaluates a constant number of candidate blocks during the disparity search, simplifying the memory-access pattern and the processing-unit design. The number of candidate blocks evaluated by the UDS algorithm can be calculated using the Eqs. (5.1)–(5.4). The First step is performed with the number of candidate blocks denoted by $\#CB_{S1}$ in Eq. (5.1). The Search Range (SR) can be modified, however, SR = 64 was used, as defined in the 3D-HTM default configuration. This means that candidate blocks located 64 samples to the left and 63 samples to the right of the collocated block can be compared in a search using the UDS algorithm. In Eq. (5.2), $\#CB_{S2}$ represents the candidate blocks evaluated in the Second Step, whereas in Eq. (5.3) $\#CB_{S3}$ denotes the candidate blocks compared in the Horizontal Refinement step. Equation (5.4) is used to calculate the total number of candidate blocks.

$$\#CB_{S1} = \frac{2 \times SR}{S1} + 1 \tag{5.1}$$

$$\#CB_{S2} = 2 \times \left(\frac{S1}{S2} - 1 \right) \tag{5.2}$$

$$\#CB_{S3} = 2 \times (S2 - 1) \tag{5.3}$$

$$\#CB_{Total} = CB_{S1} + CB_{S2} + CB_{S3} \tag{5.4}$$

In order to evaluate the UDS algorithm and to define the S_1 and S_2 values, some experiments were performed based on the CTC document recommended by JCT-3V. The experiments were conducted using the 3D-HTM reference software (3D-HEVC Reference Software 2024) and the RA temporal configuration. The eight 3D video sequences and the three views (texture + depth map) of these video sequences were used along with the four recommended QP sets defined in the CTC document (see Appendix 1). The SR remains the same defined in the 3D-HTM default configuration (64 samples).

Table 5.11 Number of candidate blocks evaluated by the UDS algorithm according to the operation point

UDS steps	Number of candidate blocks according to the operation point (S_1–S_2)	
	4–2	12–4
First step	33	11
Second step	2	4
Horizontal refinement	2	6
Total	37	21

The evaluations performed to assess the UDS performance considered the following different values of S_1 and S_2 parameters: $S_1 = \{4,8,12,16,20,24,28,32\}$ and $S_2 = \{2,4,8,12,16\}$. These evaluations were done without any modification in the 3D-HEVC quadtree structure, i.e., DE prediction can adopt any of all 24 possible PU sizes. The only change was related to the disparity estimation, which uses UDS instead TZS for both texture and depth-map DE. This evaluation was anchored in two important variables: (i) the impact in coding efficiency, measured using the BD-Rate metric, and (ii) the number of candidate blocks, which strongly impacts in throughput, power and memory traffic.

Twenty-one test cases were applied to the UDS algorithm using a notation (S_1-S_2). The BD-Rate increase was evaluated considering the average values for all videos. The lowest impact in coding efficiency was observed for case 4–2, with a BD-Rate increase of 0.4037%. However, case 4–2 is also among those cases that evaluate more candidate blocks (37). In turn, the 12–4 case is one of those cases that evaluate fewer candidate blocks (21), and it presents the lowest impact regarding compression efficiency among those cases that evaluate 21 candidate blocks, with a BD-Rate increase of 1.7288%. Based on the results, these two test cases were adopted as the operation points in the UDS algorithm: Operation Point (4–2), Operation Point (12–4).

Table 5.11 summarizes the number of candidate blocks evaluated (based on Eqs. (5.1)–(5.4)) for each step of the UDS algorithm according to these two operation points.

5.2.1.2   IUDS: Improved Unidirectional Disparity Search Algorithm

The UDS was developed to perform three steps of subsampled FS only in the horizontal direction. The three steps are similar, however, each step has a different subsampling parameter, which represents the distance that each candidate block will be apart from the neighboring candidate blocks. To get advantages of some vertical displacements exploration, which are not supported by UDS, the HDS was proposed (see Sect. 5.1.3.2) in which performs a vertical refinement to the UDS heuristic. For that, the third step from UDS has been changed, and this step also evaluates the six vertical candidates around the best horizontal candidate results from the second step. Candidates at vertical positions have a low probability to be selected to represent the disparity, however, to cover little vertical displacements it is important to assure good matching under slight cameras vibrations or captured noise.

The HDS algorithm proposed in this book implements a vertical refinement, but both horizontal distances that each candidate block will be apart from the neighboring candidate blocks and the refinement vertical are proposed empirically. The iUDS algorithm brings a proper analysis to verify the impact of this approach in video coding. Therefore, to propose the iUDS algorithm (Improved UDS), three vertical refinement alternatives along with the UDS original operation points were evaluated.

All experiments to evaluate the iUDS were performed using the 3D-HTM in version 16.2 (3D-HEVC Reference Software 2024), considering the 3D-HEVC CTC document. The results presented in this Section consider the average of all videos and QPs defined in the CTC (see Appendix A) running under the RA temporal configuration. Furthermore, it is worth mentioning that between 200 and 300 AUs were encoded from each video sequence. The SR remains the same defined in the 3D-HTM default configuration (64 samples).

The iUDS vertical refinement is treated as a fourth UDS step, where the algorithm can evaluate two, four or six additional neighboring blocks in the vertical direction (three above and three below the current best block), as presented in Fig. 5.14. The vertical-refinement patterns are called here: (i) Vertical; (ii) Diagonal; (iii) Full Refinement (Vertical + Diagonal). Vertical pattern selects the two blocks located directly above and below the best matching block of the previous steps for comparison. Diagonal pattern selects for comparison the four blocks located diagonally from the best matching block. Full-refinement pattern selects both the Vertical and Diagonal blocks, totalizing six vertical neighboring blocks of the best matching block chose in the previous steps. The definition of which mode will be used is made by the N_4 parameter. After this step, the iUDS completes its execution, and the result is the block most similar to the block being encoded among the evaluated blocks.

Based on the proposed vertical-refinement patterns, a set of evaluations were performed to assess the iUDS performance when the iUDS replaces the TZS for both texture and

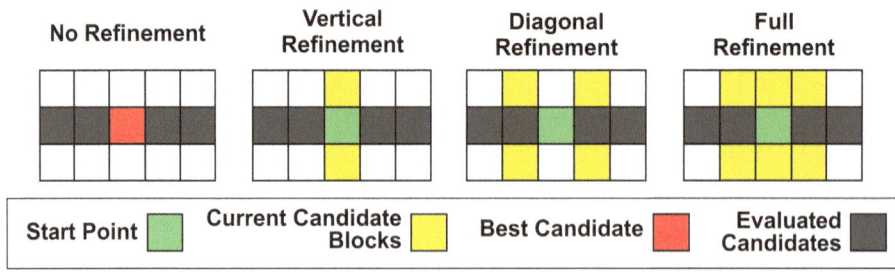

Fig. 5.14 iUDS vertical refinement analysis: refinement patters (Adapted from Perleberg et al. (2020a))

depth-map DE. This evaluation was anchored in two important variables: (i) the impact in coding efficiency, measured using the BD-Rate metric, and (ii) the number of candidate blocks, which strongly impacts in throughput, power, and memory access. Figure 5.15 presents a scatter graph containing eight test cases (TC) applied to the iUDS algorithm (black shapes). Red shapes denote the original UDS operation points, i.e., iUDS with no refinement. The BD-Rate increase is presented considering the average values for all videos of the experimental setup. The lowest impact in coding efficiency was observed for TC_4 operating with Full Refinement, with a BD-Rate increase of 0.2363%. However, TC_4 is also the case that evaluates more candidate blocks (43). In turn, TC_7 with Diagonal Refinement presents a better balance between evaluated candidate blocks (25) and BD-Rate increase (0.6374%). The complete BD-Rate results, according to the 3D-video sequences, can be observed in Appendix C.

Comparing the iUDS approach with the TZS algorithm, iUDS has a low impact on the coding efficiency and delivers an expressive reduction in the number of evaluated blocks. In (Afonso et al. 2019a), the number of candidate blocks evaluated by the TZS was analyzed. TZS evaluates 55 candidate blocks on average (standard deviation equal to 149) in the DE process, ranging from 10 up to 1,484 in the performed experiments. As one can notice in Fig. 5.15, iUDS algorithm evaluates between 21 and 43 candidate blocks for the eight TCs, and it causes a reduction between 21.8% and 61.8% in the number of evaluated blocks, considering the TZS average results, whereas presenting a predictable behavior in terms of computational effort.

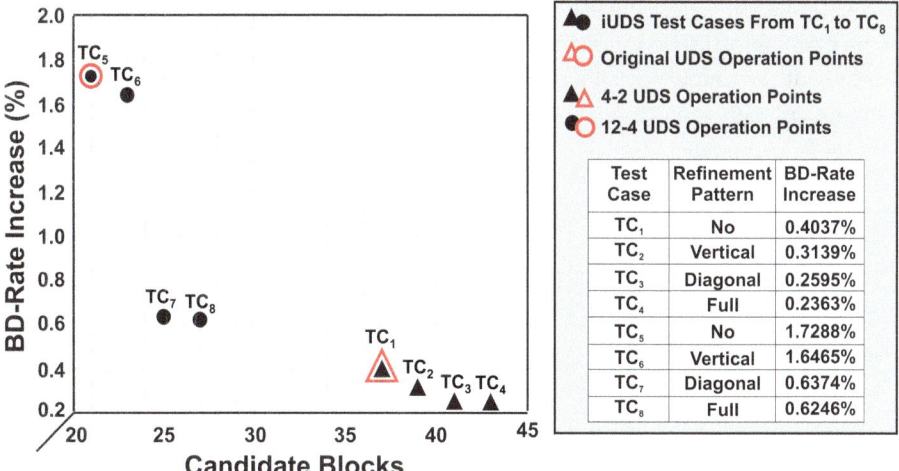

Fig. 5.15 iUDS vertical refinement analysis: scatter graph containing eight iUDS test cases (Adapted from Perleberg et al. (2020a))

5.2.2 IUDS 3D-HEVC DE Architecture

The architecture designed for the iUDS algorithm is presented in Fig. 5.16, and it supports the eight test cases presented in Sect. 5.2.1.2. Also, it can process the 24 possible PU sizes with 24 specialized datapaths for each PU size (red dotted lines in Fig. 5.16). All datapaths work in parallel ensuring the processing of high-resolution 3D videos. Each specialized datapath has a variable number of internal modules, and each module is an instance of the iUDS, developed to efficiently deal with only one PU size. The number of modules to process each PU size is based on the PU width, where each datapath has the number of modules needed to parallelize all the PUs of a CTU with 64×64 samples, as given by Eq. (5.5). So, the number of iterations for each module completes the processing of a CTU depends on the PU height, as given by Eq. (5.6).

$$Modules_{Datapath} = \frac{CTU_{Width}}{Width_{Datapath}} \tag{5.5}$$

$$Iterations_{Module} = \frac{CTU_{Height}}{Height_{Datapath}} \tag{5.6}$$

As presented in Fig. 5.16, each module of this architecture can be divided into two units. The first unit is the control unit, which selects the candidate blocks using any subsampling parameter. The control unit also compares the results provided by different steps of iUDS. The second unit is the Search and Comparison Unit (SCU), which compares the two candidate blocks selected by the control unit with the current block following the SAD similarity criterion. The SAD was adopted as the similarity criterion because it can be easily implemented in hardware, and it is the similarity criterion most commonly used in hardware implementations (Fan et al. 2018; Perleberg et al. 2018). Also, it is the similarity criterion adopted in the 3D-HTM to deal with the DE step, by default.

5.2.2.1 Search and Comparison Unit

The block diagram of one SCU is also presented in Fig. 5.16. Targeting the best tradeoff between processing rate and the idle time of the circuit, each SCU processes, in parallel, two candidate blocks so that one line of each candidate block is processed at each clock cycle, independently of the block size. For that, each SCU was composed of two SAD Trees, two Accumulators and one Comparator of two inputs, as presented in Fig. 5.16.

Each SAD Tree is responsible for calculating the SAD value between one line of one reference block and one line of the current block. Figure 5.16 exemplifies the SAD Tree that deals with PUs with 8-sample width. One can notice that each SAD Tree receives the samples of each line of these candidate blocks and calculates the difference between them and the samples of the current block. After that, the absolute value of that difference is obtained. Finally, these absolute differences between the samples are added two by two, so that each SAD Tree results in only one SAD value. These sums are separated into three

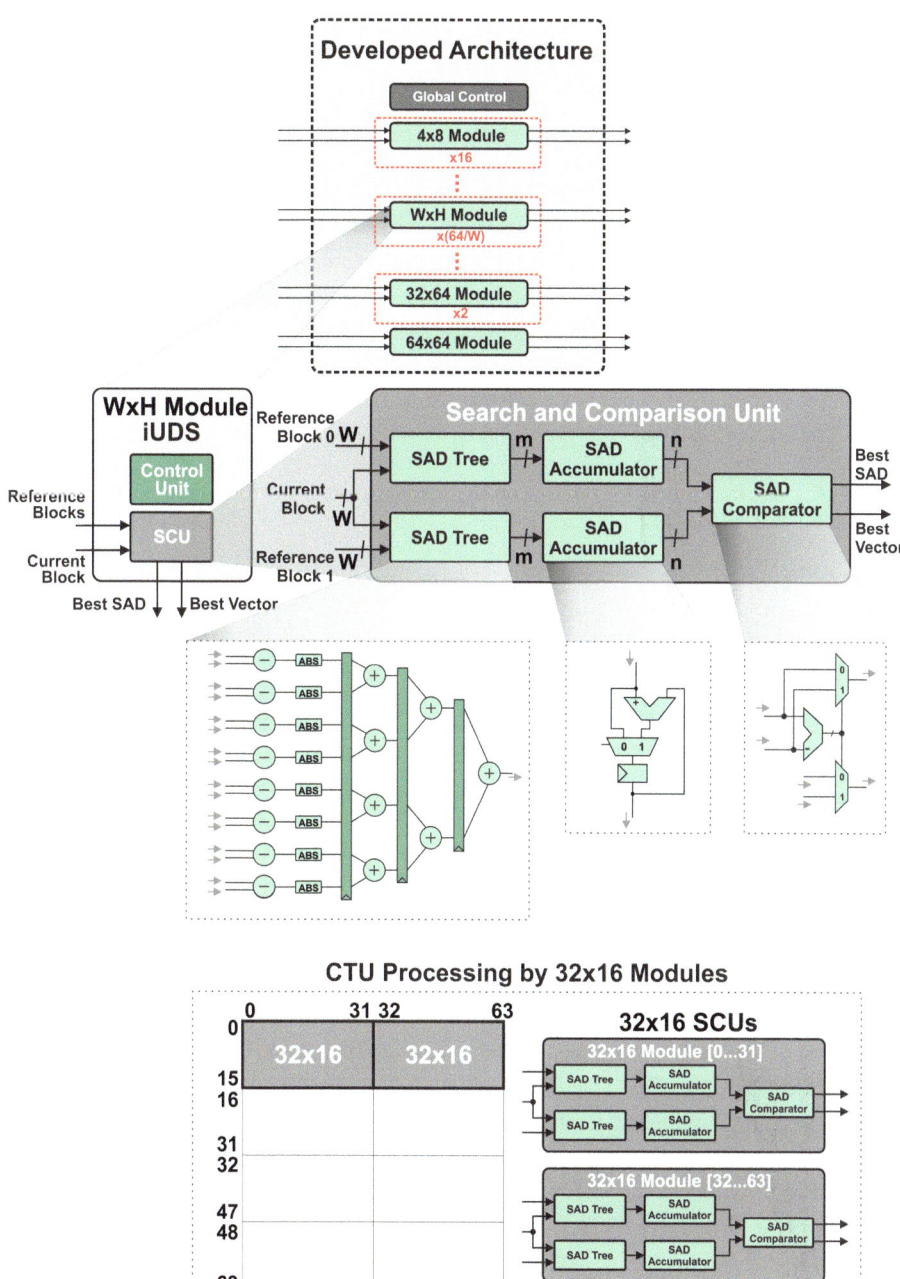

Fig. 5.16 iUDS developed architecture (Adapted from Perleberg et al. (2020a))

pipeline stages. Since the SAD Trees of other PU sizes will have a different number of inputs, they will also require a different number of pipeline stages to maintain the desired throughput.

After the SAD Trees compute the SAD value related to one line, the processed SAD values are accumulated by two SAD Accumulators, while the SAD Trees compute the rest of the block lines. After the SAD Trees compute all lines of the candidate blocks, the output value in each Accumulator is exactly the SAD value between each candidate block and the current block. Figure 5.16 shows the architecture of one SAD Accumulator. The Control signal is responsible for resetting the value accumulated by the SAD Accumulator when the SAD Trees start the processing of a new block. The value of **m** (input sample bit depth) depends on the processed PU width, while the value of **n** (output bit depth) depends on the width and height of the processed PU. The values of **m** and **n** are given by Eqs. (5.7) and (5.8), respectively.

$$m = 8 + \log_2(PU_{WIDTH}) \tag{5.7}$$

$$n = m + \log_2(PU_{HEIGHT}) \tag{5.8}$$

Finally, when the SAD Trees finish the blocks processing, the Comparator stage processes the SAD values along with the disparity vectors of each block and delivers the smallest SAD value and the related motion vector. The architecture of a SAD Comparator is presented in Fig. 5.16. The most significant bit (MSB) of the difference between the two SAD values is used as the selector in two multiplexers that output the smallest SAD and its disparity vector.

5.2.2.2 Control Unit

The control unit selects the candidates to be processed by the SCU. The selection is made from left to right, starting with the candidates on the left side of the SA, and finishing by the candidates on the right side of the SA.

The number of candidate blocks that the first three iUDS steps need to select for comparison is dependent on the subsampling parameters of the adopted operation point, as it can be obtained by Eq. (5.9), where **N#** is the subsampling parameter from the iUDS step. However, the number of candidates from the vertical refinement are 2, 4, or 6 for the Vertical, Diagonal, or Full Refinement, respectively. The control unit selects the candidate blocks two by two, from the left of the SA to the right of the SA, and send the selected blocks to the SCU. All results from SCU will be compared by the control unit. So, when the search covers the SA, the search can pass to the processing of the next step candidates.

$$h(N_\#) = \left(SA_{Width_\#} - CTU_{Width} + N_\#\right)\Big/N_\# \tag{5.9}$$

The control unit is also responsible for managing the memory communication. The memory communication affects the processing of each CTU, which will need a few more clock cycles to be processed. Two additional cycles are required at the end of the processing of each PU, and one additional cycle is required when ending the process of each CTU.

5.2.2.3 Temporal Analysis

The number of clock cycles that a module of a specific datapath needs to process one specific iUDS step is given by Eq. (5.10), being **h** the number of candidates from the respective step. The equation considers the filling of the pipeline stages in the first iteration of each iUDS step, the processing rate of one block line per clock cycle, and also the two clock cycles due to the memory communication. When considering the four iUDS steps, the number of clock cycles to process each block can be obtained by the Eq. (5.11).

$$p(N_\#) = log_2(PU_{Width}) + \frac{h(N_\#)}{2}(PU_{Height}) + 1 + 1 \qquad (5.10)$$

$$r(N_1, N_2, N_3, N_4) = p(N_1) + p(N_2) + p(N_3) + p(N_4) \qquad (5.11)$$

As previously mentioned, some modules were replicated targeting a minimum idle time to each module, instead of using only one module per PU size to process the CTU. This choice resulted in a balanced number of clock cycles, as it can be confirmed using the Eq. (5.12), that represents the total clock cycles that each module needs to fully complete the processing of a CTU.

$$s(N_1, N_2, N_3, N_4) = \left(\frac{CTU_{Height}}{PU_{Height}}\right)(r(N_1, N_2, N_3, N_4) + 2) + 1 \qquad (5.12)$$

As one can notice from Eq. (5.12), the modules which process the 8×4 datapath are the modules that need more clock cycles to process a CTU, considering any iUDS operation point among the eight test cases evaluated in Sect. 5.2.1.2. So, the delay of this module is used to determine the throughput of the developed hardware. The slower modules that have finished the processing of the CTU were disabled using the clock-gating technique, reducing energy consumption.

The clock cycle diagram of one WxH module is presented in Fig. 5.17. As one can notice, the processing of each PU depends on the block size and the subsampling parameters. Similarly, the number of iterations from a module to process a given PU depends on the block Height.

5.2.3 Synthesis Results and Comparisons

The developed hardware was synthesized targeting an ASIC technology according to the experimental setup presented in Appendix B. The results are presented in Table 5.12, along with hardware results of the main related works. Although the two related works

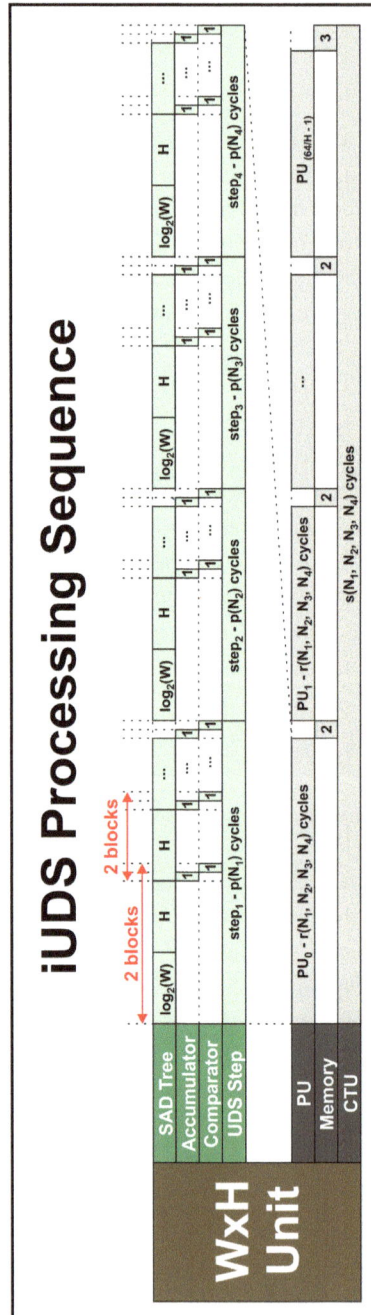

Fig. 5.17 Temporal analysis of the iUDS developed architecture (Adapted from Perleberg et al. (2020a))

Table 5.12 Hardware design synthesis results

Related works		Fan (2018)	Perleberg (2018)	This book	
				TC$_5$	TC$_4$
Target encoding tool		HEVC ME	HEVC ME	3D-HEVC DE	
Block sizes		All 24 possible	Four	All 24 possible	
BMA algorithm		Modified TZS	Modified TZS	iUDS	
ASIC technology		65 nm	Nangate 45nm	Nangate 45nm	
Total area (gates)		489.4 k	18,103k	3,243k	
SRAM		18.4kB	no	16.77kB*	
1080p@60fps 1 view	Frequency	500 MHz	144.92MHz	58.8MHz	105.8 MHz
	Power/voltage	128.5mW/ N.A	140.84mW/ N.A	47.82mW/ 0.95V	90.72mW/ 0.95V
3D-video processing at UHD 2160p@60fps		No (0.25 views)	No (1.76 views)	Yes (3.4 views)	

*Estimated

target the ME step, some values related to the hardware results (such as the power dissipation and area usage) can be compared, since both ME and DE are similar encoding tools. Other comparisons, as the compression efficiency obtained with the BMA developed by (Fan et al. 2018; Perleberg et al. 2018) and BMA proposed in this book, cannot be made since they were evaluated in a different encoding context, using different encoders, with different video sequences. Furthermore, as our hardware uses the MVD format which requires processing of both texture and depth maps, the throughput of related works was divided by two.

As presented in Table 5.12, iUDS architecture needs more resources than (Fan et al. 2018) and less than (Perleberg et al. 2018). Except (Fan et al. 2018), that used 65nm standard cells, all other works used 45nm. However, the results show that our design reaches power dissipation ranging from 29 to 62% smaller than (Fan et al. 2018), and from 35 to 66% smaller than (Perleberg et al. 2018). The power results related to the developed architecture considered the Test Cases TC$_4$ and TC$_5$, which represent the best operation points considering coding efficiency and computationally effort, respectively. Besides, the designed hardware reaches throughput to process UHD 2160p 3D videos at 60fps (3.4 views) while these two related works did not reach this throughput.

The replacement of the TZS by the iUDS can bring better hardware results when compared to TZS related works. Also, this replacement affects the on-chip memory characteristics. The iUDS needs to process candidate blocks from only three lines considering a CTU, reducing the samples stored in the on-chip memory. Considering the CTU size

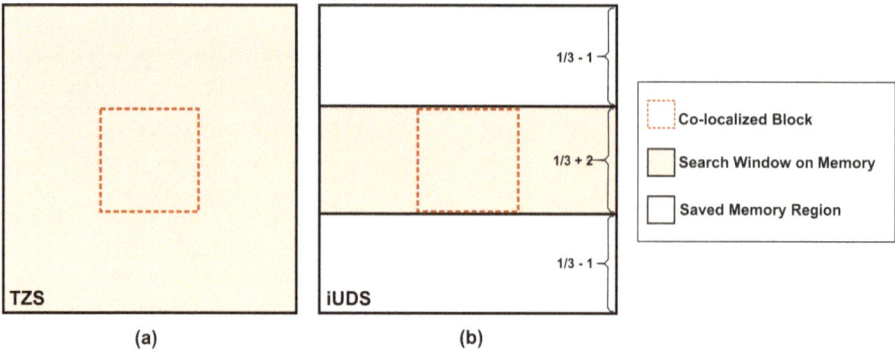

Fig. 5.18 On-chip memory required to store the search window for one CTU: **a** Using TZS algorithm; **b** using the iUDS algorithm (Adapted from Afonso et al. (2018))

of 64×64 samples, and an SR of 64 samples, the iUDS needs an SA of 192×66 samples, which represents a memory of 16.77kB. This value represents an on-chip memory reduction of 65.62% since the storage of all TZS candidates requires a memory of 192×192 samples. Figure 5.18 helps to understand this reduction, where a search area (a.k.a. Search Window—SW) with the size of 192×192 samples (necessary to employ 64×64 CTU-level data reuse with search range of 64 samples) considering the TZS algorithm is reduced to an SW of 192×66 samples when using the iUDS algorithm. Additionally, after processing the search algorithm for a given block, the on-chip memory must be refreshed with the SW of the new block to be predicted. Considering a search window-based data reuse strategy, such as Level-C (Chen et al. 2006), this fact would increases in three the total number of memory accesses for the TZS algorithm, and in one additional access considering the iUDS algorithm in anyone operation point.

The on-chip memory reduction is important to allow a hardware area reduction, but mainly to allow lower energy consumption, as discussed in the following.

An estimation of the impact in terms of energy consumption on memory-related issues when the iUDS algorithm replaces the TZS algorithm on the 3D-HEVC disparity-estimation prediction is presented in Table 5.13 presents a comparison between the test case TC4 (4–2 UDS with full vertical refinement), test case TC5 (12–4 UDS without refinement), and TZS algorithms regarding energy consumption of on-chip memory estimated by looking at the characteristics of these algorithms, such as the number of candidate blocks and search range. Note that these results were normalized concerning TZS algorithm since it evaluates more candidate blocks among the algorithms presented in Table 5.13.

One may notice in Fig. 5.18 that TZS requires an on-chip memory almost three times larger than that of iUDS to store the SW. This fact indicates that the static energy consumption tends to be almost three times greater when employing TZS. Actually, the

Table 5.13 Estimated normalized energy consumption

Algorithm	#Cand blocks	Estimated normalized energy consumption		
		Leakage	Dynamic	SAD operations
TZS	55	1	1	1
TC4 iUDS (4–2 UDS with full vertical refinement)	37	0.344	0.759	0.782
TC5 iUDS (12–4 UDS without refinement)	21	0.344	0.379	0.382

iUDS leakage energy consumption is 65.62% lower than in TZS according to the energy estimation by considering an iUDS operation with full vertical refinement.

Based on these values, the normalized dynamic energy consumption of reading/writing operations was estimated for TZS (1), TC4 iUDS (0.759), and TC5 iUDS (0.379). Note that iUDS propitiates a reduction of dynamic energy from 24.1% up to 62.1% in comparison with TZS. It is worth mentioning that TZS has a dynamic number of evaluated candidates, that depends on the video content and the prediction tool evaluated. Considering the DE context, the TZS evaluates 55 candidates (Afonso et al. 2019a) on average. However, the iUDS evaluates from 21 to 43 candidates, which represents a reduction from 21.81 to 61.81% in the number of evaluated candidates. Considering 130 candidates for this search algorithm (average + 0.5 standard deviation), the dynamic energy consumption would be 6.04 times higher than that consumed by TC5 iUDS. Moreover, considering the worst TZS case (1,484 candidates), the memory energy consumption would be 67.59 times higher than that consumed by TC5 iUDS in this scenario.

The relationship between dynamic (reading and writing) and static (leakage) energy consumption varies according to the technology. This fact hampers fair estimative regarding total energy consumption. Still, one may notice that the hardware implementation of the iUDS algorithm would strongly reduce the energy consumption and size of on-chip memory to store the disparity SW. Since TZS requires the evaluation of more candidate blocks than iUDS, it demands higher parallelism considering a hardware implementation to maintain the same architecture throughput. Note that increasing the architecture parallelism will lead to greater energy consumption by DE processing units.

Therefore, this dedicated hardware design for the 3D-HEVC Disparity Estimation presented a high-throughput and low-power, using the proposed iUDS algorithm. A software analysis was performed to investigate the optimal operation points considering a vertical refinement step, allowing the selection of the operation points from an external control signal according to the desired trade-off between coding efficiency and power dissipation. The synthesis results showed that the developed hardware obtains higher throughput than the related works using the TZS algorithm. Also, by using the iUDS instead of TZS, the architecture reduces in 65.62% the required memory size. Finally, the designed architecture was the first dedicated hardware design for 3D-HEVC DE step that supports all 24 possible PU sizes capable of processing UHD2160p 3D-videos at 60 fps.

Conclusions

The popularization of multimedia services has pushed forward the development of video-capable embedded mobile devices. In this scenario, immersive video systems capable of providing high-quality video-related experience based on the simultaneous exhibition of multiple views are also expected, including systems capable of dealing with high and ultra-high resolutions. To meet this demand, the prominent technology is called multi-view plus depth (MVD), which can produce a realistic immersive experience by generating synthetic video views on the decoder side, reducing the amount of information on the encoder side. This is important because the MVD approach allows an expressive reduction in the amount of information that must be captured, encoded, and transmitted, when compared with previous approaches. The MVD approach was first explored in a video coding standard at the HEVC (High Efficiency Video Coding) extension called 3D-HEVC. The MVD approach associates a depth-map with each texture frame for each view that composes the video sequence. With the texture and depth maps, the decoder is able to generate additional texture synthesized views allowing an improved immersive experience. Considering this new scenario, with multiple texture and depth map pictures, novel encoding tools were defined to improve the coding efficiency of MVD when compared with traditional 2D video encoding. These new coding tools increased the coding efficiency but also increased the computational effort required to compress the video content. Since video-capable embedded mobile devices require efficient energy/memory-management strategies to deal with severe memory/processing requirements and limited energy supply, the development of high-throughput and energy-efficient systems targeting the MVD encoding is essential. This book exploited the MVD characteristics to propose high-throughput and energy-efficient architectures/systems focusing on the 3D-HEVC. Along with this book, an extensive discussion on the current state-of-the-art

© The Author(s), under exclusive license to Springer Nature Switzerland AG 2025   177
V. Afonso et al., *Hardware Design for 3D Video Coding*, Synthesis Lectures on
Engineering, Science, and Technology, https://doi.org/10.1007/978-3-031-80232-4_6

3D video encoding was also provided. Afterward, background knowledge was presented followed by our main contributions which were divided into three main topics: (i) 3D-HEVC reference software evaluations; (ii) Intra-frame prediction architectures; and (iii) Inter-prediction architectures.

Six architectures/systems exploiting memory/processing aspects, and the characteristics of the encoding tools under an MVD approach were developed, four of them focused on the Intra-frame prediction tools and the other two focused on the Inter prediction, which includes both Inter-frames and Inter-view predictions. All the decisions taken during the development of the architectures were marked in exhaustive simulations using the 3D-HEVC reference software where the following data were extracted: (i) the most time demanding encoding tools used for both texture and depth-map coding according to the 3D-HEVC to allow the understanding of its energy requirements; (ii) the most selected and representative 3D-HEVC encoding tools used during the coding process to identify the tools with more impact regarding compression and image quality; (iii) the impact regarding compression by constraining specific 3D-HEVC encoding tools and the block sizes supported by these encoding tools during the coding process.

All the developed architectures take advantage of application-specific knowledge of 3D-HEVC, i.e., its new coding tools and video content properties.

Also, on-/off-chip memory-related access behaviors related to the 3D-HEVC encoding process were analyzed at run time to adapt the memory management and to save energy consumption for the ME/DE system developed in this book. This system also takes advantage of a flexible coding order between texture and depth maps to propose heuristics and memory management capable of reducing the energy consumption of its video memory and processing architectures.

The low-power hardware design for the Depth Intra Skip coding tool reduces the computational effort based on a strategy that consisted of replacing the SVDC for the SAD as the similarity criterion. This strategy reduced the number of arithmetic operations related to the similarity criterion by over 71%, and it avoided a rendering process. This architecture was the first dedicated hardware design published in the literature for the DIS tool, and it is capable of processing five UHD 2160p views at 60 frames per second.

The low-power and memory-aware depth-map Intra-frame prediction system was developed based on hardware-oriented heuristics to reduce the computational effort. These strategies consisted of removing less important prediction modes and block sizes. Also, a specific complexity-reduction strategy was applied to the DMM-1 along with a modified Bresenham implementation to represent the wedgelet at the time, which allowed the calculation of only six wedgelet patterns without any storage prediction. This architecture was the first to support both the novel 3D-HEVC and the conventional HEVC intra prediction tools, and it is capable of processing HD 1080p nine views at 30 fps.

The presented 6WR algorithm and its dedicated high-throughput and energy-aware hardware design for the 3D-HEVC DMM-1 used the Bresenham algorithm to avoid memory requirements. The 6WR algorithm exploited the edge gradients, reducing 98.5% of the evaluated wedgelets and 97.62% in the DMM-1 runtime, with an impact between 1.2 and 2.8% in the coding efficiency. The synthesis results show that the 6WR architecture can process up to nine HD1080p views at 30 fps, the highest performance in comparison with the related works. Considering the same performance, the 6WR architecture has a lower impact on the encoding efficiency and the best result in energy-efficiency between related works.

The complete intra-frame prediction hardware design for the 3D-HEVC based on a flexible coding order between texture and depth channels reduced the computational effort and, consequently, the power dissipation. A set of hardware-friendly heuristics was proposed and implemented in the 3D-HTM. The smallest computational effort reduction was reached when adopting only the IPHOC, which skips the evaluation of the 4×4 PBs and each PB size evaluates only 33 wedgelets in the DMM-1 mode. In this case, an average time reduction of 6.23% was reached at the cost of a BD-Rate increase of 2.57% in RA temporal configuration. The highest computational effort reduction was reached when adopting the LC-ICDSD heuristic, which includes the IPHOC and ICDSD heuristics. When using LC-ICDSD, the computational effort was reduced by 8.35%, at the cost of a 5.35% BD-Rate increase. The developed design dissipates 384.6 mW when processing three views of HD 1080p videos at 30 fps (three texture views and their associated depth maps) considering the LC-ICDSD heuristic. This architecture was the first solution that was able to process 3D-HEVC intra-frame prediction that supported FCO. In addition, this was the first architecture to cover both texture and depth map processing, including luminance and chrominance, for the 3D-HEVC intra-frame prediction.

The Motion and Disparity Estimation system was designed for low-energy consumption, featuring a run-time adaptive memory hierarchy. The memory hierarchy featured window-based prefetching, data reuse, subsampling, and dynamic voltage scaling controlled by the developed Depth-Based Dynamic Search Window Resizing algorithm. The developed energy-aware Motion and Disparity Estimation system was the first solution published in the literature that proposed a real-time ME/DE system for the 3D-HEVC standard with an adaptive memory hierarchy capable of processing three HD 1080p views at 30 frames per second. Memory results demonstrated an average on-chip energy reduction of 79% in comparison to the Level-C solution.

The low-power and coding-efficient disparity estimation architecture based on the developed Improved Unidirectional Disparity-Search algorithm (iUDS) prioritized the horizontal search instead of using conventional search algorithms in two dimensions to take advantage of the fact that horizontal disparities are more common. Results showed negligible impact on the encoding efficiency and enough throughput to encode five UHD 2160p views at 40 frames per second.

The proposed algorithms and architectures widely exploited 3D-HEVC memory/processing aspects, and the characteristics of the encoding tools under an MVD approach. Two of the developed systems also took advantage of a flexible coding order between texture and depth maps to reduce the energy consumption. One of them focuses on the

Motion and Disparity Estimations, i.e., focusing on the Inter-frames and Inter-view predictions, and the other one extends this approach to the Intra-frame prediction, covering both the Intra tools inherited by the HEVC standard as the ones introduced by the 3D-HEVC. This way, the decisions used for the coding process of a specific channel can be used to avoid or simplify the coding process of the other channel. An example of this idea is related to the angular mode inherited by the HEVC and the DMM-1 mode introduced by the 3D-HEVC. Both encoding tools are based on the prediction of directional structures, which assumes that these modes can also take advantage of sharing their decisions.

Appendix A: 3D-HTM Evaluations Experimental Setup

Adopted Color Space

The color space consists of a system able to represent the colors in an image (Ohm 2015). The color space YCbCr is composed of three components: (i) Luminance, denoted by Y; (ii) Chrominance blue, denoted by Cb; and (iii) Chrominance red, denoted by Cr (Ohm 2015). This color space is one of the main systems used in video coding (Ohm 2015) since the brightness information (luminance channel) is separated from the color information (chrominance channels) and then, the different channels can be independently processed without losses in the visual quality. This way, the encoders take advantage of the human visual system which is more sensitive to the brightness information than the color information to reduce the amount of data used to represent the image (Ohm 2015). For example, video encoders apply a technique called color subsampling, which consists of reducing the spatial sampling of the chrominance information when compared to the luminance, increasing the coding efficiency (Ohm 2015).

The color subsampling can be applied with different spatial subsampling, which defines the color subsampling formats 4:4:4, 4:2:2, and 4:2:0. In the format 4:4:4, there are four samples of Cb and four samples of Cr for every four samples of Y. The format 4:2:2 considers two samples of Cb and two samples of Cr for every four samples of Y. Finally, the format 4:2:0 uses only one sample of Cb and one sample of Cr for every four samples of Y. The format 4:4:4 maintains the same spatial sampling for both luminance and chrominance samples and, therefore, it does not change the image quality but also it does not provide any compression. By using the format 4:2:0, the image quality decreases but the total video size is reduced to the half without the use of any other compression technique. The quality losses are almost imperceptible for the human visual system, even with this dramatic discard of data. By default, the HEVC standard uses the format 4:2:0, but the formats 4:4:4 and 4:2:2 can also be used when configured for that. The HEVC also allows

V. Afonso et al., *Hardware Design for 3D Video Coding*, Synthesis Lectures on Engineering, Science, and Technology, https://doi.org/10.1007/978-3-031-80232-4

for disabling the use of chrominance samples (monochromatic option) (Bross et al. 2012). In this book, we adopted the format used by default in the HEVC, i.e., the format 4:2:0.

Adopted Coding Efficiency Metric

There are several metrics in the literature targeting the coding efficiency assessment of digital videos. These metrics use two well-known main approaches that consist of: (i) a subjective assessment; (ii) an objective assessment. In summary, the metrics based on a subjective assessment take into account the perception of the spectators of both the original and the modified video in a given set of test conditions while the objective approach uses equations to compare the original and the modified video.

Considering the objective metrics, such as the bit rate and the PSNR, the tradeoff between compression and image quality cannot be accurately verified when they are separately analyzed. This way, the metrics BD bit rate (or BD-BR or BD-Rate) and BD Peak Signal-to-Noise Ratio (BD-PSNR), based on the BjØntegaard Difference (BD) (Bjɸntegaard 2001), are widely adopted in the video-coding research area. The BD-Rate measures the percentage variation of the bit rate between two test cases considering encoded videos with the same objective image quality. Positive values of BD-Rate indicate low efficiency on the compression since it means a bit-rate increase is needed to obtain the same image quality. In the BD-PSNR metric, the idea is similar, where the percentage variation of PSNR values (in decibels) between two test cases is evaluated considering the same bit rate. This way, negative values of BD-PSNR indicate an image degradation to obtain the same bit rate. The test cases should be encoded considering the Common Test Conditions (Müller et al. 2014), which are presented in Appendix A.

To obtain the BD-PSNR and the BD-Rate values between two test cases (here called Reference and Tested), these test cases are encoded with four different Quantization Parameters (QPs), which results in eight pairs (PSNR, bit rate). From these eight pairs, two rate-distortion (RD) curves are generated with a third-order interpolation function, as presented in Fig. A.1. The area between these two RD curves passes through an integration that considers the y-axis as the reference for BD-Rate calculation (Fig. A.1a), and considers the x-axis as the reference for BD-PSNR calculation (Fig. A.1b). Equations (A.1) and (A.2) can be used to calculate the BD-Rate and the BD-PSNR, respectively. REF and TEST in (A.1) and (A.2) represent PSNR or bit rate values for the curves related to the test case Reference and Tested, respectively. The variables a and b represent the second minimum and the second maximum PSNR values from both curves for the BD-BR calculation, or they represent the second minimum and the second maximum bit-rate values from both curves for the BD-PSNR calculation, respectively. In this book, the metrics BD-BR and BD-PSNR were adopted to measure the coding efficiency of the proposed algorithms and architectures.

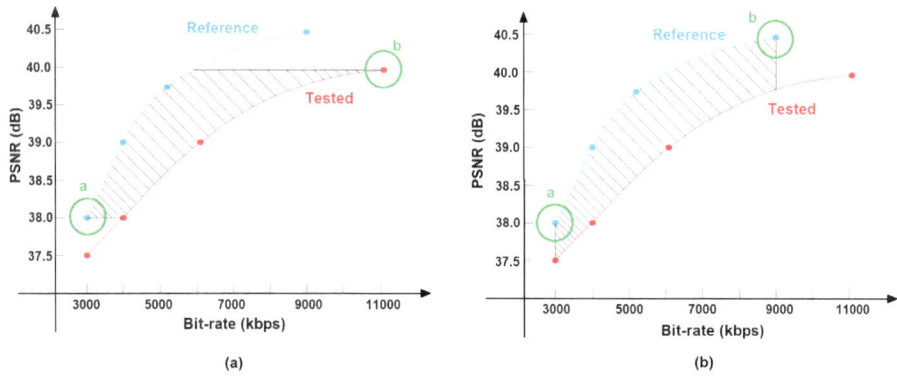

Fig. A.1 RD curves used to **a** BD-BR and **b** BD-PSNR calculations

$$BD_{BR} = \frac{\int_a^b (REF_{PSNR}(y) - TEST_{PSNR}(y))dy}{b - a} \tag{A.1}$$

$$BD_{PSNR} = \frac{\int_a^b (REF_{BR}(x) - TEST_{BR}(x))dx}{b - a} \tag{A.2}$$

Test Conditions

The experimental conditions of the evaluations performed and presented in this book are based on documents available by the JCT-3V group. In the document JCT3V-G1100 (Müller et al. 2014), the JCT-3V recommended test conditions to conduct experiments with the 3D-HTM reference software. Among these conditions, the document defines eight 3D-video sequences using MVD format to be used, as presented in Table A.1. In Table A.1, it is also possible to observe the views to be used as inputs in the experiments using each video sequence, given these 3D-videos have a different number of views. This way, any researcher can conduct the experiments under the same conditions as the others.

The configurations and test conditions provided by the JCT-3V recommend the use of an Inter-View encoding structure composed of three views with a coding order center/left/right and P-I-P Inter-View prediction. The JCT-3V document also recommends the use of a temporal prediction structure with GoP containing eight frames and Intra-frame occurrences every 24 frames (Random Access every second). Also, the document defines the use of inputs with 8-bit wide samples, internal processing with 8-bit wide samples, and complete resolution for both texture and depth maps.

Table A.1 Recommended 3D-video sequences and the views to be used in experiments (Müller et al. 2014)

ID	Sequence	Views to be used (left-center-right)
S01	Poznan_Hall2	7-6-5
S02	Poznan_Street	5-4-3
S03	Undo_Dancer	1-5-9
S04	GT_Fly	9-5-1
S05	Kendo	1-3-5
S06	Balloons	1-3-5
S08	Newspaper	2-4-6
S10	Shark	1-5-9

The QP values to be used in the experiments are also defined in the JCT-3V document. The QP values defined by the JCT-3V to be used in the Texture Independent Views are 25, 30, 35, and 40. The QP values to be used in the depth maps are fixed according to the texture QP values, as presented in Table A.2. QP_{V0} represents the QP values for the texture whereas QP_{D0} represents the QP values for their respective depth maps. Therefore, for the texture QP values recommended by the JCT-3V the correspondent depth-map QP values are 34, 39, 42, and 45 (in bold in Table A.2).

The eight 3D-video sequences with MVD format recommended by the JCT-3V have between 200 and 300 AUs so that five of them are in the 1920×1088 resolution and the other three are in the 1024×768 resolution, as presented in Table A.3. The frame rate of the sequences varies between 25 and 30 frames per second (fps), which results in 10 s of video, except for the sequence *Poznan_Hall2*. Five sequences are real sequences (S01, S02, S05, S06, and S08) recorded using seven or nines cameras spaced every five or 13.75 cm whereas the other three sequences are artificially generated using computer graphics (S03, S04, and S10), as shown in Table A.3.

The main image aspects related to each one of the sequences are presented in Table A.4. Note that the sequences provided by the JCT-3V present well-diversified characteristics: sequences with internal recording environment, external recording environment, or computer graphics; sequences with or without camera movement; sequences with different levels of object movement; sequences with natural or artificial lighting; as well as different image details and depth structures.

The first frames of the 3D-video test sequences provided by the JCT-3V can be observed in Figs. A.2 and A.3, in which are presented the texture pictures and their respective depth maps. Figure A.2 shows the 1920×1088 sequences, whereas Fig. A.3 shows the 1024×768 sequences.

Several architectures were developed based on heuristics and techniques evaluated with exhaustive software simulations. Once these software evaluations occurred at different moments, the experimental setup adopted for each one of the experiments had slight variations. These variations occurred because new versions of the 3D-HEVC reference software and standardization documents emerged during the development of the

Table A.2 QP values to be used for texture and depth maps (Müller et al. 2014)

QP_{V0}	51	50	49	48	47	46	45	44	43	42	41	**40**	39	38	37	36	**35**	34	33	32	31	**30**	29	28	27	26	**25**
QP_{D0}	51	50	50	50	50	49	48	47	46	45	45	**45**	44	44	43	43	**42**	42	41	41	40	**39**	38	37	36	35	**34**

Table A.3 Technical aspects of the 3D-video sequences

Test sequence	Resolution	AUs	Frame rate (fps)	Cameras	Inter-lens camera spacing
S01—Poznan_Hall2	1920 × 1088	200	25	9	13,75 cm
S02—Poznan_Street	1920 × 1088	250	25	9	13,75 cm
S03—Undo_Dancer	1920 × 1088	250	25	9	–
S04—GT_Fly	1920 × 1088	250	25	9	–
S05—Kendo	1024 × 768	300	30	7	5 cm
S06—Balloons	1024 × 768	300	30	7	5 cm
S08—Newspaper1	1024 × 768	300	30	9	5 cm
S10—Shark	1920 × 1088	300	30	9	–

Table A.4 Main image aspects of the 3D-video sequences

Test sequence	Recording environment	Object movement	Camera movement	Image detailing level	Depth structures	Light
S01—Poznan_Hall2	Internal	Complex (reflections and transparences)	Yes	Intermediate	Complex	Artificial
S02—Poznan_Street	External	Complex (reflections and transparences)	No	High	Complex	Natural
S03—Undo_Dancer	Computer Graphics	Complex	–	High	Simple	Computer graphics
S04—GT_Fly	Computer Graphics	Simple	–	Intermediate	Complex	Computer graphics
S05—Kendo	Internal	Complex (smoke, reflections and transparences)	Yes	High	Intermediate	Artificial studio light
S06—Balloons	Internal	Complex (reflections and transparences)	Yes	Intermediate	Intermediate	Artificial studio light
S08—Newspaper	Internal	Simple	No	High	Intermediate	Artificial studio light
S10—Shark	Computer graphics	Complex	–	High	Complex	Computer graphics

architectures. Anyway, the absolute majority of the experiments followed the recommendations given in the common test conditions, i.e., they used all eight video sequences and four QP pairs recommended by the JCT-3V, as presented in Table A.5. Also, almost all experiments used the 3D-HTM in version 16 and considered all AUs of the sequences. Table A.5 depicts the exact experimental setup used to guide each evaluation done using the 3D-HTM.

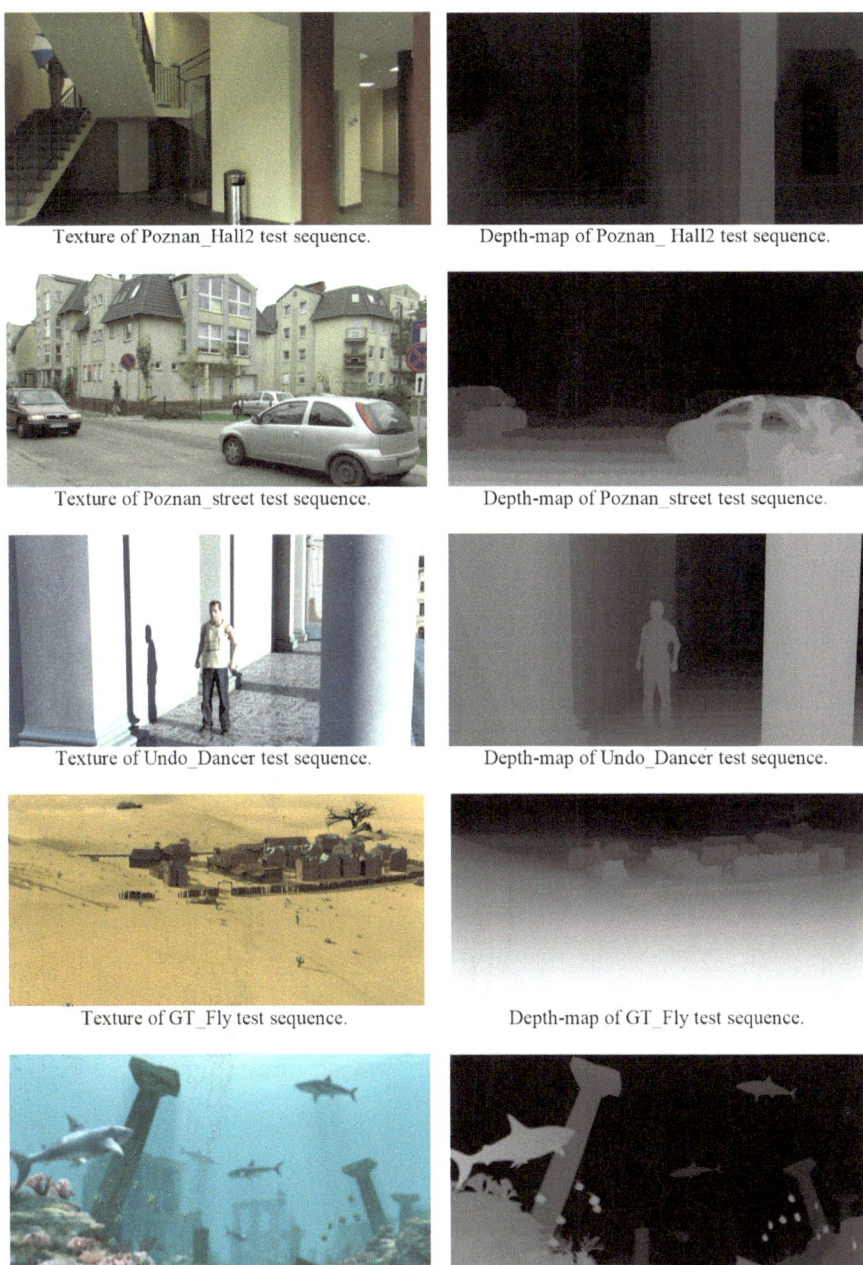

Texture of Poznan_Hall2 test sequence.   Depth-map of Poznan_ Hall2 test sequence.

Texture of Poznan_street test sequence.   Depth-map of Poznan_street test sequence.

Texture of Undo_Dancer test sequence.   Depth-map of Undo_Dancer test sequence.

Texture of GT_Fly test sequence.   Depth-map of GT_Fly test sequence.

Texture of Shark test sequence.   Depth-map of Shark test sequence.

Fig. A.2 First frames of the 1920 × 1088 3D-video test sequences (Müller et al. 2014)

Texture of Balloons test sequence.

Depth-map of Balloons test sequence.

Texture of Newspaper1 test sequence.

Depth-map of Newspaper1 test sequence.

Texture of Kendo test sequence.

Depth-map of Kendo test sequence.

Fig. A.3 First frames of the 1024×768 3D-video test sequences (Müller et al. 2014)

Table A.5 Experimental setup summary according to each experiment

Evaluations	3D-HEVC reference software version	Temporal configuration	Coding order configuration	Sequences	Quantization parameters	Number of access units
Section 3.1.1	3D-HTM 16.0	All intra	Conventional coding order	Eight (according to the CTC)	Four pairs (according to the CTC)	All AUs (between 200 and 300 AUs)
Section 3.1.2	3D-HTM 16.0	Random access	Conventional coding order	Two sequences (ballons and Undo_Dancer) recommended in the CTC	One QP pair recommended in the CTC (30/39)	All AUs (between 200 and 300 AUs)
Section 3.2	3D-HTM 16.0	Random access	Conventional and flexible coding orders	Eight (according to the CTC)	Four pairs (according to the CTC)	All AUs (between 200 and 300 AUs)
Section 4.1	3D-HTM 16.0	All intra	Conventional coding order	Eight (according to the CTC)	Four pairs (according to the CTC)	All AUs (between 200 and 300 AUs)
Section 4.2	3D-HTM 15.1	All intra and random access	Conventional coding order	Eight (according to the CTC)	Four sets (according to the CTC)	All AUs (between 200 and 300 AUs)
Section 4.3	3D-HTM 16.2	All intra and random-access	Conventional coding order	Eight (according to the CTC)	Four pairs (according to the CTC)	All AUs (between 200 and 300 AUs)
Section 4.4	3D-HTM 16.2	All intra and random-access	Flexible coding order	Eight (according to the CTC)	Four pairs (according to the CTC)	All AUs (between 200 and 300 AUs)

(continued)

Table A.5 (continued)

Evaluations	3D-HEVC reference software version	Temporal configuration	Coding order configuration	Sequences	Quantization parameters	Number of access units
Section 5.1	3D-HTM 16.2	Random access	Flexible coding order	Eight (according to the CTC)	Four pairs (according to the CTC)	48 AUs
Section 5.2	3D-HTM 16.2	Random access	Conventional coding order	Eight (according to the CTC)	Four pairs (according to the CTC)	All AUs (between 200 and 300 AUs)

Appendix B: Synthesis Methodology Applied to the Developed Architectures/Systems

All developed architectures and systems were described in VHDL, and the synthesis results were generated targeting an ASIC technology using the Cadence Encounter RTL Compiler 11.10 tool (Cadence 2024). The ASIC synthesis results were obtained for 45 nm Nangate standard cells (Nangate 2024), except for the architectures presented in Sects. 4.3 and 4.4 where the ASIC synthesis results were obtained for 40 nm TSMC standard cells (TSMC 2024). Also, the gate count presented in this book is calculated based on 2-input NANDs.

The power results of all developed architectures were also estimated using the Cadence RTL Compiler 11.10 tool. These power results were provided by RTL Files and default switching activity targeting 45 nm standard-cells and they considered a voltage supply of 0.95 V. All developed architectures used the clock-gating low-power technique enabled for the automatic logic synthesis.

The default flow for the logic synthesis provided by the RTL Compiler tool is basically composed of four steps, as follows: (i) Setup, responsible for setting the target library and activating some improvements such as clock gating; (ii) Elaborate Design, responsible for identifying the structure and building the architecture; (iii) Set Timing and Design Constraints, which defines timing and operating conditions for the design; (iv) Synthesize Design step, responsible for the design synthesis and mapping to the target technology, which generates the gate level net list of standard cells available in the library (Nangate 2024). After these steps, the power analysis can be done, which is responsible for analyzing and reporting the power estimation.

Figure B.1 depicts the default flow provided by the RTL Compiler tool for power estimation. It requires, as inputs the RTL files (HDL files), and the Synopsys Design Constraints file (SDC file), containing the operating conditions for the design, such as target frequency, input and output delay, clock latency, etc. This flow considers a default switching-activity value, defined by the RTL Compiler tool. The default toggle percentage is 20% on each input pin, where the toggle percentage is a multiplication factor (scale factor—0.2) to be used with the clock to modify the toggle rate.

© The Editor(s) (if applicable) and The Author(s), under exclusive license to Springer Nature Switzerland AG 2025

V. Afonso et al., *Hardware Design for 3D Video Coding*, Synthesis Lectures on Engineering, Science, and Technology, https://doi.org/10.1007/978-3-031-80232-4

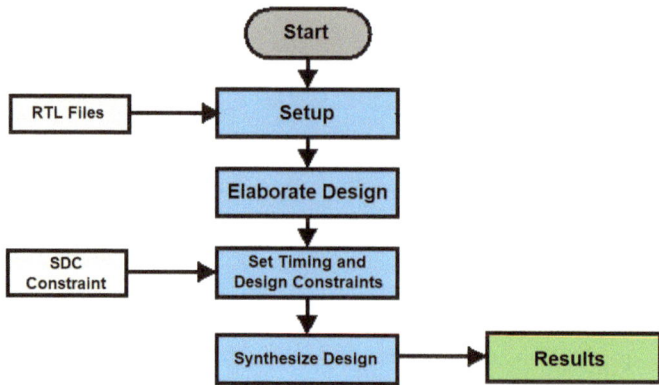

Fig. B.1 Default flow for power estimation provided by the RTL compiler tool

Appendix C: BD-Rate Results According to the 3D-Video Sequences

This appendix shows the impact of the compression according to the 3D-video sequences for the four architectures/systems presented in this book. The results consider the BD-Rate metric. When *Video Total* is referred, the quality related to texture channel (disregarding synthesized frames) considering the total bit rate of the video (texture + depth) is measured. When *Synthesized Frames* are referred, the quality of frames synthesized with encoded texture and encoded depth maps, using DIBR, is measured in comparison with frames synthesized with original texture and original depth maps. Finally, when *Video Only* is referred, the quality related to the texture channel using the texture bit rate of the video (disregarding synthesized frames) is measured. The BD-Rate results are shown for each architecture/system divided into six subsections, as follows.

Low-Power Depth Intra Skip Architecture Based on Distortion Metric Replacing

This subsection presents the BD-Rate increase according to the 3D-video sequences for the Low-Power Depth Intra Skip Architecture presented in Chap. 4. Table C.1 presents the BD-Rate increase of replacing the SVDC with the SAD as the similarity criterion.

Low-Power and Memory-Aware Depth-Map Intra-Frame Prediction System Based on Complexity Reduction Heuristics

This subsection presents the BD-Rate variation according to the 3D-video sequences for the Low-Power and Memory-Aware Depth-map Intra-frame Prediction System based on Complexity Reduction Heuristics presented in Chap. 4. Table C.2 presents the BD-Rate variation of combining all complexity-reduction strategies considering the AI configuration. Table C.3 also shows the BD-Rate variation of combining all these strategies but considering the RA configuration. Also, Tables C.2 and C.3 present the total and depth time reductions by adopting all complexity-reduction heuristics.

© The Editor(s) (if applicable) and The Author(s), under exclusive license   193
to Springer Nature Switzerland AG 2025
V. Afonso et al., *Hardware Design for 3D Video Coding*, Synthesis Lectures on
Engineering, Science, and Technology, https://doi.org/10.1007/978-3-031-80232-4

Table C.1 BD-Rate increase according to the 3D-video sequences of replacing the SVDC by the SAD as the similarity criterion

Sequence	Video 0 (%)	Video 1 (%)	Video 2 (%)	Video only (%)	Video total (%)	Synthesized (%)
Balloons	0.00	0.00	0.00	0.00	0.16	0.23
Kendo	0.00	0.00	0.00	0.00	0.22	0.34
Newspaper_CC	0.00	0.00	0.00	0.00	0.22	0.26
GT_Fly	0.00	0.00	0.00	0.00	0.14	0.20
Poznan_Hall2	0.00	0.00	0.00	0.00	0.40	0.29
Poznan_Street	0.00	0.00	0.00	0.00	0.10	0.14
Undo_Dancer	0.00	0.00	0.00	0.00	0.07	0.02
Shark	0.00	0.00	0.00	0.00	0.17	0.21
1024×768	0.00	0.00	0.00	0.00	0.20	0.28
1920×1088	0.00	0.00	0.00	0.00	0.18	0.17
Average	0.00	0.00	0.00	0.00	0.19	0.21

Table C.2 BD-rate variation according to the 3D-video sequences in AI configuration

Sequence	Video 0 (%)	Video 1 (%)	Video 2 (%)	Video only (%)	Video total (%)	Synthesized (%)	Time reduction (%)	Depth reduction (%)
Balloons	0.00	0.00	0.00	0.00	−0.02	7.96	45.86	53.26
Kendo	0.00	0.00	0.00	0.00	0.02	8.04	46.29	53.81
Newspaper_CC	0.00	0.00	0.00	0.00	−0.89	12.09	48.35	54.46
GT_Fly	0.00	0.00	0.00	0.00	−0.61	3.82	46.68	54.01
Poznan_Hall2	0.00	0.00	0.00	0.00	−0.29	7.98	46.39	54.69
Poznan_Street	0.00	0.00	0.00	0.00	−0.39	3.31	47.27	53.88
Undo_Dancer	0.00	0.00	0.00	0.00	−0.18	7.55	45.20	53.45
Shark	0.00	0.00	0.00	0.00	−0.87	6.58	46.64	54.30
1024×768	0.00	0.00	0.00	0.00	−0.30	9.36	46.83	53.84
1920×1088	0.00	0.00	0.00	0.00	−0.47	5.85	46.44	54.07
Average	0.00	0.00	0.00	0.00	−0.41	7.16	46.58	53.98

Table C.3 BD-rate variation according to the 3D-video sequences in RA configuration

Sequence	Video 0 (%)	Video 1 (%)	Video 2 (%)	Video only (%)	Video total (%)	Synthesized (%)	Time reduction (%)	Depth reduction (%)
Balloons	0.00	0.15	0.32	0.08	0.22	2.73	14.95	32.21
Kendo	0.00	0.12	0.11	0.04	0.61	2.31	14.88	31.13
Newspaper_CC	0.00	0.21	0.16	0.06	−0.32	6.48	18.01	35.03
GT_Fly	0.00	0.75	0.54	0.13	−0.05	0.88	15.48	31.92
Poznan_Hall2	0.00	0.08	0.05	0.00	−0.15	2.32	17.01	33.94
Poznan_Street	0.00	0.57	0.71	0.18	0.00	2.13	18.57	35.61
Undo_Dancer	0.00	1.30	1.23	0.33	0.53	2.28	14.56	31.35
Shark	0.00	1.02	1.04	0.22	0.23	1.98	15.89	31.34
1024×768	0.00	0.16	0.20	0.06	0.17	3.84	15.94	32.79
1920×1088	0.00	0.74	0.72	0.17	0.11	1.92	16.30	32.83
Average	0.00	0.52	0.52	0.13	0.14	2.64	16.17	32.82

DMM-1 Energy-Aware and High-Throughput Hardware Design Based on 6WR Algorithm

This subsection presents the BD-Rate variation according to the 3D-video sequences for the Energy-Aware and High-Throughput Hardware Design Based on the 6WR Algorithm presented in Chap. 4. Tables C.4 and C.5 present the BD-Rate variation by using the 6WR DMM-1 Algorithm in AI configuration and RA configuration, repectively.

Quality-Power Configurable Flexible Coding Order Hardware Design for Real-Time 3D-HEVC Intra Frame Prediction

This subsection presents the BD-Rate variation according to the 3D-video sequences for the Quality-Power Configurable Flexible Coding Order Hardware Design for Real-Time 3D-HEVC Intra Frame Prediction presented in Chap. 4. Tables C.6, C.7, C.8, C.9, C.10 and C.11 from present the BD-Rate variation by using IPHOC, or IPHOC + ICDSD, or IPHOC + ICDSD + LC-ICDSD heuristics in AI configuration and RA configuration.

Table C.4 BD-rate variation according to the 3D-video sequences by using the 6WR DMM-1 algorithm in AI configuration

Sequence	Video 0 (%)	Video 1 (%)	Video 2 (%)	Video only (%)	Video total (%)	Synthesized (%)
Balloons	0.0000	0.0000	0.0000	0.0000	−0.0270	2.9502
Kendo	0.0000	0.0000	0.0000	0.0000	−0.0443	3.1037
Newspaper_CC	0.0000	0.0000	0.0000	0.0000	−0.2689	4.8831
GT_Fly	0.0000	0.0000	0.0000	0.0000	−0.1683	1.5374
Poznan_Hall2	0.0000	0.0000	0.0000	0.0000	−0.1151	3.6767
Poznan_Street	0.0000	0.0000	0.0000	0.0000	−0.1589	1.2846
Undo_Dancer	0.0000	0.0000	0.0000	0.0000	0.0491	2.6116
Shark	0.0000	0.0000	0.0000	0.0000	−0.1138	2.5018
1024×768	0.0000	0.0000	0.0000	0.0000	−0.1134	3.6457
1920×1088	0.0000	0.0000	0.0000	0.0000	−0.1014	2.3224
Average	0.0000	0.0000	0.0000	0.0000	−0.1059	2.8187

Table C.5 BD-rate variation according to the 3D-video sequences by using the 6WR DMM-1 algorithm in RA configuration

Sequence	Video 0 (%)	Video 1 (%)	Video 2 (%)	Video only (%)	Video total (%)	Synthesized (%)
Balloons	0.0000	−0.0220	0.2055	0.0279	−0.0879	1.0506
Kendo	0.0000	0.1371	0.0430	0.0344	0.0441	0.9370
Newspaper_CC	0.0000	0.1402	−0.1442	−0.0116	−0.3623	2.6141
GT_Fly	0.0000	0.4091	0.2494	0.0716	0.0129	0.4525
Poznan_Hall2	0.0000	−0.1535	−0.1227	−0.0528	−0.1795	1.2424
Poznan_Street	0.0000	0.0771	0.3157	0.0484	−0.1151	0.7576
Undo_Dancer	0.0000	0.6897	0.7426	0.1890	0.3159	1.0746
Shark	0.0000	0.6707	0.6433	0.1398	0.1686	1.1640
1024×768	0.0000	0.0851	0.0348	0.0169	−0.1353	1.5339
1920×1088	0.0000	0.3386	0.3657	0.0792	0.0406	0.9382
Average	0.0000	0.2435	0.2416	0.0558	−0.0254	1.1616

Table C.6 BD-rate variation according to the 3D-video sequences by using IPHOC in AI configuration

Sequence	Video 0 (%)	Video 1 (%)	Video 2 (%)	Video only (%)	Video total (%)	Synthesized (%)	Time reduction (%)
Balloons	0.7745	0.9240	0.9044	0.8684	0.6317	6.1137	47.5529
Kendo	0.7718	0.8158	0.8413	0.8101	0.5846	5.9573	48.1198
Newspaper_CC	1.1018	1.2684	1.0531	1.1373	0.2232	8.8113	45.6699
GT_Fly	4.0378	4.0590	4.0369	4.0446	3.4339	7.2737	48.7984
Poznan_Hall2	0.8383	0.7634	0.7064	0.7697	0.7100	6.1358	49.5223
Poznan_Street	1.1345	1.0728	1.1668	1.1251	0.7652	3.2003	47.1706
Undo_Dancer	2.0735	2.0341	2.0997	2.0691	1.8744	7.9730	48.9914
Shark	1.4446	1.4458	1.4411	1.4438	0.4383	7.5439	49.2168
1024×768	0.8827	1.0027	0.9329	0.9386	0.4798	6.9608	47.1142
1920×1088	1.9057	1.8750	1.8902	1.8905	1.4444	6.4253	48.7399
Average	1.5221	1.5479	1.5312	1.5335	1.0827	6.6261	48.1302

Energy-Aware Motion and Disparity Estimation System with Run-Time Adaptive Memory Hierarchy

This subsection presents the BD-Rate increase according to the 3D-video sequences by adopting the proposed hardware-oriented constraints (HCs) and the developed algorithms/techniques for the Energy-Aware Motion and Disparity Estimation System presented in Chap. 5. Tables C.12, C.13, C.14, C.15, C.16 and C.17 present the results for the six scenarios tested with the HC1. Tables C.18, C.19 and C.20 present the BD-Rate increase for the three scenarios tested with the HC2. Table C.21 presents the results for the HC3. Table C.22 shows the BD-Rate increase for the HC4. Tables C.23, C.24, C.25 and C.26 present the impact on compression by using the depth subsampling considering four different test scenarios. Finally, Table C.27 shows the BD-Rate increase of combining all adopted algorithms/techniques along with all hardware-oriented constraints (HC4).

Table C.7 BD-rate variation according to the 3D-video sequences by using IPHOC in RA configuration

Sequence	Video 0 (%)	Video 1 (%)	Video 2 (%)	Video only (%)	Video total (%)	Synthesized (%)	Time reduction (%)
Balloons	0.5228	0.7067	0.6295	0.5193	0.2267	2.2778	5.5569
Kendo	0.4461	0.5691	0.0689	0.3688	0.2523	1.6739	5.8031
Newspaper_CC	1.0485	0.5711	0.7529	0.7304	−0.0778	4.7262	5.4770
GT_Fly	1.6223	2.0346	2.1707	1.5981	1.5548	2.8999	6.5122
Poznan_Hall2	0.7260	0.7784	0.6190	0.6517	0.5687	2.1561	6.2923
Poznan_Street	0.9957	0.9047	1.0125	0.8458	0.4530	1.9382	7.3427
Undo_Dancer	1.3191	2.0289	2.3899	1.3959	1.5296	2.7839	6.3088
Shark	0.5744	1.6052	2.1566	0.7463	0.4942	2.1605	6.2701
1024 × 768	0.6725	0.6156	0.4838	0.5395	0.1338	2.8926	5.6123
1920 × 1088	1.0475	1.4704	1.6697	1.0476	0.9201	2.3877	6.5452
Average	0.9069	1.1498	1.2250	0.8570	0.6252	2.5771	6.1954

Table C.8 BD-rate variation according to the 3D-video sequences by using IPHOC + ICDSD in AI configuration

Sequence	Video 0 (%)	Video 1 (%)	Video 2 (%)	Video only (%)	Video total (%)	Synthesized (%)	Time reduction (%)
Balloons	0.7745	6.4797	6.1180	4.4792	4.4076	10.2528	28.0528
Kendo	0.7718	6.9887	6.9834	4.9293	4.8854	10.7367	29.7441
Newspaper_CC	1.1018	5.0209	5.3411	3.8772	3.1437	13.0110	24.5325
GT_Fly	4.0378	10.5638	10.5143	8.3857	7.8440	13.1737	26.7589
Poznan_Hall2	0.8383	17.3395	15.7972	11.2101	10.7929	17.4776	25.1576
Poznan_Street	1.1345	8.6489	7.6951	5.8734	5.4184	8.1789	20.7466
Undo_Dancer	2.0735	6.5281	6.4793	5.0331	4.8511	14.8042	28.5339
Shark	1.4446	6.1144	6.0411	4.5361	3.9481	14.4019	25.4404
1024 × 768	0.8827	6.1631	6.1475	4.4286	4.1456	11.3335	27.4431
1920 × 1088	1.9057	9.8389	9.3054	7.0077	6.5709	13.6073	25.3275
Average	1.5221	8.4605	8.1212	6.0405	5.6614	12.7546	26.1209

Table C.9 BD-rate variation according to the 3D-video sequences by using IPHOC + ICDSD in RA configuration

Sequence	Video 0 (%)	Video 1 (%)	Video 2 (%)	Video only (%)	Video total (%)	Synthesized (%)	Time reduction (%)
Balloons	0.5228	0.8510	0.9650	0.6016	0.6127	2.8982	2.6906
Kendo	0.4461	0.7971	0.6940	0.5405	0.8679	2.4015	3.1766
Newspaper_CC	1.0485	1.2575	1.4101	0.9449	0.6992	6.0394	3.3438
GT_Fly	1.6223	2.5719	2.7514	1.6987	1.8351	4.0332	4.2205
Poznan_Hall2	0.7260	2.9270	1.9437	1.2621	1.1542	3.0106	2.9582
Poznan_Street	0.9957	0.8924	1.0883	0.8479	0.7287	2.4628	2.8270
Undo_Dancer	1.3191	3.0823	3.3211	1.6387	2.2763	5.3451	13.7413
Shark	0.5744	2.3863	3.0252	0.9043	1.6237	5.0033	2.3850
1024 × 768	0.6725	0.9686	1.0230	0.6957	0.7266	3.7797	3.0703
1920 × 1088	1.0475	2.3720	2.4259	1.2704	1.5236	3.9710	5.2264
Average	0.9069	1.8457	1.8998	1.0548	1.2247	3.8993	4.4179

Table C.10 BD-rate variation according to the 3D-video sequences by using (IPHOC + ICDSD + LC-ICDSD) in AI configuration

Sequence	Video 0 (%)	Video 1 (%)	Video 2 (%)	Video only (%)	Video total (%)	Synthesized (%)	Time reduction (%)
Balloons	0.7745	33.9859	34.2518	22.6568	22.2796	28.2449	61.7694
Kendo	0.7718	41.7775	42.9312	28.0157	27.1350	33.0472	61.6805
Newspaper_CC	1.1018	36.5412	31.3941	22.9840	21.3803	34.2779	63.9886
GT_Fly	4.0378	47.2620	47.3538	32.3062	31.0296	34.7646	62.4545
Poznan_Hall2	0.8383	52.5222	51.0874	33.8445	32.6976	41.1500	61.6970
Poznan_Street	1.1345	38.3684	38.6297	25.9160	24.9243	27.3336	61.1792
Undo_Dancer	2.0735	21.5801	22.1801	15.1932	14.8906	28.0617	67.1336
Shark	1.4446	24.0784	24.2705	16.3736	15.4309	26.5806	61.7994
1024 × 768	0.8827	37.4349	36.1923	24.5521	23.5983	31.8567	62.4795
1920 × 1088	1.9057	36.7622	36.7043	24.7267	23.7946	31.5781	62.8527
Average	1.5221	37.0145	36.5123	24.6612	23.7210	31.6826	62.7128

Table C.11 BD-Rate variation according to the 3D-video sequences by using (IPHOC + ICDSD + LC-ICDSD) in RA configuration

Sequence	Video 0 (%)	Video 1 (%)	Video 2 (%)	Video only (%)	Video total (%)	Synthesized (%)	Time reduction (%)
Balloons	0.5228	1.5346	2.5394	1.0431	1.4448	4.2699	7.1040
Kendo	0.4461	1.2631	1.7236	0.8388	1.4405	3.3987	7.6207
Newspaper_CC	1.0485	4.0641	3.8533	2.0660	2.5120	8.9972	8.1159
GT_Fly	1.6223	3.4908	3.7008	1.8940	2.4586	4.9315	8.3021
Poznan_Hall2	0.7260	4.6187	2.8988	1.7893	1.8864	4.3778	5.1270
Poznan_Street	0.9957	1.5754	2.0180	1.0937	1.3908	3.6336	6.3756
Undo_Dancer	1.3191	5.0665	4.7917	2.0852	3.2466	6.8063	17.1395
Shark	0.5744	3.5263	4.2133	1.1255	2.4738	6.3943	7.4002
1024 × 768	0.6725	2.2873	2.7054	1.3160	1.7991	5.5553	7.6135
1920 × 1088	1.0475	3.6555	3.5245	1.5975	2.2912	5.2287	8.8689
Average	0.9069	3.1424	3.2174	1.4919	2.1067	5.3512	8.3981

Table C.12 BD-rate increase according to the 3D-video sequences for the HC1 by constraining the block sizes to 8 × 8 and 16 × 16

Sequence	Video 0 (%)	Video 1 (%)	Video 2 (%)	Video only (%)	Video total (%)	Synthesized (%)
Balloons	6.5	7.2	5.5	7.4	8.1	7.65
Kendo	9.6	14.3	9.3	11.7	13.6	12.65
Newspaper_CC	2.5	5.3	3.9	4.1	5.5	5.01
GT_Fly	13.7	7.0	5.6	14.4	15.7	15.02
Poznan_Hall2	13.8	16.9	17.4	19.0	20.9	18.92
Poznan_Street	4.7	8.9	10.7	8.6	10.5	8.48
Undo_Dancer	6.3	2.2	1.6	7.5	9.8	9.05
Shark	13.2	8.4	8.5	13.5	18.7	18.89
1024 × 768	6.2	8.9	6.2	7.8	9.1	8.44
1920 × 1088	10.4	8.7	8.8	12.6	15.1	14.07
Average	8.8	8.8	7.8	10.8	12.8	11.96

Table C.13 BD-rate increase according to the 3D-video sequences for the HC1 by constraining the block sizes to 8×8 and 32×32

Sequence	Video 0 (%)	Video 1 (%)	Video 2 (%)	Video only (%)	Video total (%)	Synthesized
Balloons	3.8	4.0	3.8	4.8	5.4	5.04
Kendo	5.9	8.2	4.4	6.8	7.4	7.00
Newspaper_CC	2.1	3.0	2.5	3.1	3.6	3.30
GT_Fly	10.7	2.7	2.9	10.8	11.5	11.08
Poznan_Hall2	8.1	9.3	10.3	11.7	12.4	10.90
Poznan_Street	3.9	3.3	5.6	6.3	7.0	5.70
Undo_Dancer	5.3	0.8	0.9	6.2	7.3	6.72
Shark	10.7	5.5	4.6	10.7	12.6	12.76
1024×768	3.9	5.1	3.6	4.9	5.4	5.12
1920×1088	7.7	4.3	4.9	9.1	10.2	9.43
Average	6.3	4.6	4.4	7.6	8.4	7.81

Table C.14 BD-rate increase according to the 3D-video sequences for the HC1 by constraining the block sizes to 8×8 and 64×64

Sequence	Video 0 (%)	Video 1 (%)	Video 2 (%)	Video only (%)	Video total (%)	Synthesized (%)
Balloons	6.5	6.8	6.1	7.7	8.0	7.59
Kendo	8.4	10.4	5.8	9.3	9.6	8.98
Newspaper_CC	3.8	5.2	2.8	4.9	5.0	4.65
GT_Fly	15.1	5.7	6.5	14.8	15.4	14.89
Poznan_Hall2	8.9	8.3	9.5	11.9	12.3	11.06
Poznan_Street	6.4	4.3	7.3	9.4	10.1	8.41
Undo_Dancer	7.2	1.4	2.3	8.4	9.1	8.32
Shark	19.2	11.0	9.5	18.8	20.2	20.05
1024×768	6.2	7.5	4.9	7.3	7.6	7.08
1920×1088	11.3	6.1	7.0	12.7	13.4	12.55
Average	9.4	6.6	6.2	10.7	11.2	10.49

Table C.15 BD-rate increase according to the 3D-video sequences for the HC1 by constraining the block sizes to 16×16 and 32×32

Sequence	Video 0 (%)	Video 1 (%)	Video 2 (%)	Video only (%)	Video total (%)	Synthesized (%)
Balloons	3.09	3.80	3.17	3.97	3.98	4.92
Kendo	4.55	7.20	3.18	5.28	5.36	5.22
Newspaper_CC	1.78	2.81	2.07	2.52	2.10	2.43
GT_Fly	12.88	2.59	1.44	12.90	12.53	12.35
Poznan_Hall2	6.82	8.77	9.23	9.92	10.40	9.12
Poznan_Street	3.02	3.31	5.46	4.99	5.12	4.34
Undo_Dancer	5.47	0.12	−0.12	6.17	6.51	5.97
Shark	9.40	4.26	3.29	9.30	10.02	10.40
1024×768	3.14	4.60	2.81	3.92	3.81	4.19
1920×1088	7.52	3.81	3.86	8.66	8.92	8.44
Average	5.88	4.11	3.46	6.88	7.00	6.85

Table C.16 BD-Rate increase according to the 3D-video sequences for the HC1 by constraining the block sizes to 16×16 and 64×64

Sequence	Video 0 (%)	Video 1 (%)	Video 2 (%)	Video only (%)	Video total (%)	Synthesized (%)
Balloons	3.5	4.6	3.3	4.7	4.7	5.34
Kendo	4.7	7.0	3.5	5.5	5.6	5.36
Newspaper_CC	2.2	3.1	2.0	3.1	2.8	3.03
GT_Fly	13.6	2.9	2.6	13.4	12.9	12.57
Poznan_Hall2	4.7	4.6	6.5	6.9	7.2	6.66
Poznan_Street	3.3	2.7	3.4	5.2	5.2	4.28
Undo_Dancer	5.7	0.2	0.1	6.7	6.6	5.96
Shark	11.1	4.4	4.0	10.8	10.8	11.05
1024×768	3.5	4.9	2.9	4.5	4.4	4.58
1920×1088	7.7	2.9	3.3	8.6	8.5	8.10
Average	6.1	3.7	3.2	7.0	7.0	6.78

Table C.17 BD-rate increase according to the 3D-video sequences for the HC1 by constraining the block sizes to 32×32 and 64×64

Sequence	Video 0 (%)	Video 1 (%)	Video 2 (%)	Video only (%)	Video total (%)	Synthesized (%)
Balloons	5.0	4.4	4.9	5.9	5.8	6.94
Kendo	7.5	8.5	4.2	7.9	7.2	7.28
Newspaper_CC	3.7	3.6	2.1	4.4	2.9	3.55
GT_Fly	31.1	5.3	5.1	29.8	27.1	26.72
Poznan_Hall2	7.5	5.8	6.3	8.9	8.5	7.74
Poznan_Street	4.9	2.0	4.7	6.9	6.5	5.56
Undo_Dancer	12.1	2.8	4.8	12.9	11.8	10.99
Shark	19.5	7.9	6.2	18.8	16.7	17.22
1024×768	5.4	5.5	3.7	6.1	5.3	5.92
1920×1088	15.0	4.7	5.4	15.5	14.1	13.64
Average	11.4	5.0	4.8	11.9	10.8	10.75

Table C.18 BD-rate increase according to the 3D-video sequences for the HC2 by removing the TZS predictors (except the *Zero Predictor*) (in addition to the HC1)

Sequence	Video 0 (%)	Video 1 (%)	Video 2 (%)	Video only (%)	Video total (%)	Synthesized (%)
Balloons	3.3	4.1	3.1	4.1	4.1	4.90
Kendo	5.3	7.7	3.8	6.1	6.5	6.29
Newspaper_CC	1.8	4.1	2.5	3.2	3.0	3.07
GT_Fly	21.4	5.0	5.6	21.1	19.8	19.41
Poznan_Hall2	7.7	9.8	12.7	11.9	11.9	10.33
Poznan_Street	8.0	4.2	5.4	10.0	9.7	8.53
Undo_Dancer	7.5	0.9	1.1	8.3	8.4	7.72
Shark	19.6	7.6	4.7	19.2	17.7	18.07
1024×768	3.5	5.3	3.1	4.5	4.5	4.75
1920×1088	12.8	5.5	5.9	14.1	13.5	12.81
Average	9.3	5.4	4.9	10.5	10.1	9.79

Table C.19 BD-rate increase according to the 3D-video sequences for the HC2 by removing the TZS predictors (except the *Zero Predictor*), and the *Raster* step (in addition to the HC1)

Sequence	Video 0 (%)	Video 1 (%)	Video 2 (%)	Video only (%)	Video total (%)	Synthesized (%)
Balloons	3.24	4.15	2.85	4.19	4.36	4.93
Kendo	5.91	7.95	4.64	6.83	7.02	6.70
Newspaper_CC	1.79	4.62	3.04	3.31	3.17	3.51
GT_Fly	29.16	7.34	6.79	28.56	26.21	25.58
Poznan_Hall2	8.55	11.14	13.57	13.21	13.45	11.18
Poznan_Street	9.16	4.16	5.99	11.08	10.56	9.43
Undo_Dancer	7.51	1.10	0.96	8.04	8.09	7.63
Shark	26.30	11.29	7.06	25.78	23.86	24.05
1024×768	3.65	5.57	3.51	4.77	4.85	5.05
1920×1088	16.14	7.01	6.87	17.33	16.44	15.57
Average	11.45	6.47	5.61	12.62	12.09	11.63

Table C.20 BD-rate increase according to the 3D-video sequences for the HC2 by removing the TZS predictors (except the *Zero Predictor*), the *Raster* step, and the *Refinement* step (in addition to the HC1)

Sequence	Video 0 (%)	Video 1 (%)	Video 2 (%)	Video only (%)	Video total (%)	Synthesized (%)
Balloons	7.0	10.4	8.2	9.8	9.5	9.35
Kendo	15.4	18.4	14.2	18.5	18.5	16.51
Newspaper_CC	3.2	12.3	18.5	8.7	8.8	7.49
GT_Fly	97.5	44.2	25.9	96.6	88.8	87.07
Poznan_Hall2	29.3	30.7	39.4	41.0	38.6	33.21
Poznan_Street	25.1	26.9	27.8	33.0	31.9	28.03
Undo_Dancer	29.2	6.5	10.3	30.5	28.8	27.93
Shark	95.1	47.5	35.1	93.3	91.4	91.57
1024×768	8.5	13.7	13.6	12.3	12.3	11.11
1920×1088	55.3	31.2	27.7	58.9	55.9	53.56
Average	37.7	24.6	22.4	41.4	39.6	37.64

Table C.21 BD-rate increase according to the 3D-video sequences for the HC3 by considering the horizontal-only disparity search (in addition to the HC2)

Sequence	Video 0 (%)	Video 1 (%)	Video 2 (%)	Video only (%)	Video total (%)	Synthesized (%)
Balloons	3.24	4.57	6.63	5.31	5.76	5.74
Kendo	5.91	8.06	11.28	8.41	9.08	8.02
Newspaper_CC	1.79	4.49	19.00	5.96	6.20	5.47
GT_Fly	29.16	8.10	5.25	28.86	26.84	26.10
Poznan_Hall2	8.55	12.86	28.57	18.29	18.81	14.45
Poznan_Street	9.16	5.02	35.45	16.93	17.11	13.20
Undo_Dancer	7.51	0.89	0.30	8.72	8.86	8.06
Shark	26.30	11.38	5.55	25.89	24.58	24.72
1024×768	3.65	5.71	12.30	6.56	7.01	6.41
1920×1088	16.14	7.65	15.02	19.74	19.24	17.31
Average	11.45	6.92	14.00	14.79	14.65	13.22

Table C.22 BD-rate increase according to the 3D-video sequences for the HC4 by considering the SAD as the only similarity criterion for ME/DE, bi-directional prediction disabling, and the search limitation to one reference frame per direction (in addition to the HC3, defined as the B-Encoder configuration and the Base Memory Hierarchy—BMH)

Sequence	Video 0 (%)	Video 1 (%)	Video 2 (%)	Video only (%)	Video total (%)	Synthesized (%)
Balloons	6.02	8.05	7.17	6.88	7.21	8.38
Kendo	22.01	18.79	17.16	22.43	21.51	19.86
Newspaper_CC	5.11	10.06	5.97	6.42	6.90	7.14
GT_Fly	46.79	18.99	19.68	45.57	42.96	42.23
Poznan_Hall2	17.32	21.14	21.94	22.80	22.89	20.08
Poznan_Street	17.27	12.96	16.66	19.50	18.99	17.53
Undo_Dancer	17.31	7.73	7.21	18.09	17.31	16.10
Shark	58.71	32.28	30.78	57.90	55.51	54.47
1024×768	11.05	12.30	10.10	11.91	11.87	11.79
1920×1088	31.48	18.62	19.25	32.77	31.53	30.08
Average	23.82	16.25	15.82	24.95	24.16	23.22

Low-Power and Coding-Efficient Disparity Estimation Architecture Based on Improved Unidirectional Disparity-Search Algorithm

This subsection presents the BD-Rate variation according to the 3D-video sequences by adopting the iUDS algorithm rather than using the TZS algorithm in the 3D-HEVC DE. This approach was integrated into the Low-power and Coding-Efficient Disparity Estimation Architecture presented in Chap. 5. Four test cases (TCs) are considered in this appendix: TC1, TC4, TC5, and TC8. The BD-Rate variation for these TCs are presented in Tables C.28, C.29, C.30 and C.31, respectively.

Table C.23 BD-rate increase according to the 3D-video sequences for the depth subsampling technique by removing the four less significant bits (in addition to the HC4, defined as the Reduced-Size Hierarchy—RSH)

Sequence	Video 0 (%)	Video 1 (%)	Video 2 (%)	Video only (%)	Video total (%)	Synthesized (%)
Balloons	6.02	7.18	7.60	6.76	6.87	8.21
Kendo	22.01	19.32	16.69	22.26	21.76	20.48
Newspaper_CC	5.11	9.58	6.06	6.38	6.81	7.31
GT_Fly	46.79	20.44	19.84	45.68	43.47	42.96
Poznan_Hall2	17.32	20.97	22.87	22.70	23.19	21.26
Poznan_Street	17.27	13.21	16.90	19.71	19.53	18.04
Undo_Dancer	17.31	8.65	7.47	18.23	17.93	17.04
Shark	58.71	32.89	31.83	57.92	56.05	55.17
1024×768	11.05	12.02	10.12	11.80	11.81	12.00
1920×1088	31.48	19.23	19.78	32.85	32.03	30.89
Average	23.82	16.53	16.16	24.95	24.45	23.81

Table C.24 BD-rate increase according to the 3D-video sequences for the depth subsampling technique by removing the four more significant bits (in addition to the HC4)

Sequence	Video 0 (%)	Video 1 (%)	Video 2 (%)	Video only (%)	Video total (%)	Synthesized (%)
Balloons	6.02	8.06	6.87	6.83	6.95	8.94
Kendo	22.01	18.58	16.90	22.32	22.43	21.61
Newspaper_CC	5.11	9.65	5.87	6.45	7.08	8.29
GT_Fly	46.79	21.34	20.22	45.70	44.62	44.49
Poznan_Hall2	17.32	21.28	22.96	22.66	23.23	21.87
Poznan_Street	17.27	13.09	16.65	19.56	19.99	18.78
Undo_Dancer	17.31	8.56	7.43	18.20	19.75	20.82
Shark	58.71	32.35	30.89	58.11	62.68	63.77
1024×768	11.05	12.10	9.88	11.87	12.15	12.95
1920×1088	31.48	19.32	19.63	32.84	34.05	33.95
Average	23.82	16.61	15.97	24.98	25.84	26.07

Table C.25 BD-rate increase according to the 3D-video sequences for the depth subsampling technique by removing the even bits (in addition to the HC4)

Sequence	Video 0 (%)	Video 1 (%)	Video 2 (%)	Video only (%)	Video total (%)	Synthesized (%)
Balloons	6.02	7.44	6.83	6.70	6.94	8.25
Kendo	22.01	18.62	17.07	22.38	21.78	20.57
Newspaper_CC	5.11	10.47	5.92	6.35	6.71	7.57
GT_Fly	46.79	20.40	19.89	45.56	43.54	42.99
Poznan_Hall2	17.32	21.43	22.23	22.84	23.36	21.27
Poznan_Street	17.27	12.58	16.87	19.53	19.71	18.45
Undo_Dancer	17.31	7.95	7.09	18.18	17.90	17.03
Shark	58.71	32.56	30.36	58.10	59.80	60.00
1024 × 768	11.05	12.18	9.94	11.81	11.81	12.13
1920 × 1088	31.48	18.98	19.29	32.84	32.86	31.95
Average	23.82	16.43	15.78	24.96	24.97	24.52

Table C.26 BD-rate increase according to the 3D-video sequences for the depth subsampling technique by removing the odd bits (in addition to the HC4)

Sequence	Video 0 (%)	Video 1 (%)	Video 2	Video only (%)	Video total (%)	Synthesized (%)
Balloons	6.02	7.87	7.16	6.76	6.91	8.35
Kendo	22.01	19.00	16.58	22.24	21.59	20.36
Newspaper_CC	5.11	10.21	6.24	6.32	6.56	7.58
GT_Fly	46.79	19.80	18.91	45.41	43.44	43.10
Poznan_Hall2	17.32	21.47	22.57	22.75	23.39	21.79
Poznan_Street	17.27	13.53	16.48	19.72	19.51	18.03
Undo_Dancer	17.31	8.56	7.69	18.18	19.19	19.23
Shark	58.71	32.79	31.44	57.96	56.84	56.27
1024 × 768	11.05	12.36	9.99	11.78	11.69	12.10
1920 × 1088	31.48	19.23	19.42	32.80	32.47	31.68
Average	23.82	16.65	15.88	24.92	24.68	24.34

Table C.27 BD-rate increase according to the 3D-video sequences using all proposed techniques (in addition to the HC4 and depth subsampling, defined as the Run-Time Adaptive Hierarchy— RAH)

Sequence	Video 0 (%)	Video 1 (%)	Video 2 (%)	Video only (%)	Video total (%)	Synthesized (%)
Balloons	6.02	6.72	7.21	6.86	7.05	8.30
Kendo	22.01	19.42	17.19	22.57	22.14	20.66
Newspaper_CC	5.11	12.58	6.46	7.00	7.47	7.74
GT_Fly	46.79	21.54	20.99	45.72	43.45	43.00
Poznan_Hall2	17.32	28.34	33.15	28.69	30.02	25.71
Poznan_Street	17.27	12.27	16.62	19.37	19.04	17.69
Undo_Dancer	17.31	8.23	7.17	18.35	18.36	17.66
Shark	58.71	32.57	31.43	57.95	57.50	57.30
1024×768	11.05	12.91	10.29	12.14	12.22	12.24
1920×1088	31.48	20.59	21.87	34.02	33.67	32.27
Average	23.82	17.71	17.53	25.81	25.63	24.76

Table C.28 BD-rate variation according to the 3D-video sequences by using the iUDS under the test case 1 (UDS 4–2 without refinement step)

Sequence	Video 0 (%)	Video 1 (%)	Video 2 (%)	Video only (%)	Video total (%)	Synthesized (%)
Balloons	0.0000	0.4814	0.3425	0.1883	0.1815	0.1360
Kendo	0.0000	0.2038	0.0117	0.0637	0.0410	0.0377
Newspaper_CC	0.0000	0.5097	0.8659	0.2989	0.2785	0.2552
GT_Fly	0.0000	0.9818	1.1340	0.4393	0.4473	0.3361
Poznan_Hall2	0.0000	3.0303	1.8482	1.1318	1.0747	0.9198
Poznan_Street	0.0000	2.4736	2.6187	0.8525	0.8382	0.7493
Undo_Dancer	0.0000	1.3111	1.1077	0.4653	0.5237	0.4672
Shark	0.0000	−0.4126	−1.1487	−0.0960	−0.1552	−0.0565
1024×768	0.0000	0.3983	0.4067	0.1836	0.1670	0.1430
1920×1088	0.0000	1.4768	1.1120	0.5586	0.5458	0.4832
Average	0.0000	1.0724	0.8475	0.4180	0.4037	0.3556

Table C.29 BD-rate variation according to the 3D-video sequences by using the iUDS under the test case 4 (UDS 4–2 + full refinement)

Sequence	Video 0 (%)	Video 1 (%)	Video 2 (%)	Video only (%)	Video total (%)	Synthesized (%)
Balloons	0.0000	0.1524	0.2596	0.0945	0.0963	0.0566
Kendo	0.0000	0.2428	−0.1538	0.0289	0.0291	0.0061
Newspaper_CC	0.0000	0.2470	0.2112	0.1164	0.1090	0.0350
GT_Fly	0.0000	1.1140	1.0396	0.3691	0.3739	0.2700
Poznan_Hall2	0.0000	1.8637	1.0671	0.6753	0.6339	0.5633
Poznan_Street	0.0000	1.0427	1.3690	0.3911	0.3846	0.3233
Undo_Dancer	0.0000	1.0488	0.8795	0.3573	0.4099	0.2868
Shark	0.0000	−0.2346	−1.0003	−0.0842	−0.1464	−0.1160
1024 × 768	0.0000	0.2141	0.1057	0.0800	0.0781	0.0325
1920 × 1088	0.0000	0.9669	0.6710	0.3417	0.3312	0.2655
Average	0.0000	0.6846	0.4590	0.2436	0.2363	0.1781

Table C.30 BD-rate variation according to the 3D-video sequences by using the iUDS under the test case 5 (UDS 12–4 without refinement step)

Sequence	Video 0 (%)	Video 1 (%)	Video 2 (%)	Video only (%)	Video total (%)	Synthesized (%)
Balloons	0.0000	2.9720	3.6236	1.3217	1.2620	0.8549
Kendo	0.0000	3.2785	2.6801	1.2071	1.1174	0.7409
Newspaper_CC	0.0000	3.7622	8.3066	2.4617	2.2788	1.7575
GT_Fly	0.0000	2.4013	2.6036	1.0351	1.0931	0.6835
Poznan_Hall2	0.0000	7.2381	5.4676	2.8249	2.6908	2.1790
Poznan_Street	0.0000	12.9342	12.7752	3.9473	3.8302	3.0250
Undo_Dancer	0.0000	2.1452	2.4020	0.9447	1.0601	0.8743
Shark	0.0000	1.0775	0.6685	0.4344	0.4983	0.3868
1024 × 768	0.0000	3.3376	4.8701	1.6635	1.5527	1.1178
1920 × 1088	0.0000	5.1592	4.7834	1.8373	1.8345	1.4297
Average	0.0000	4.4761	4.8159	1.7721	1.7288	1.3127

Table C.31 BD-rate variation according to the 3D-video sequences by using the iUDS under the test case 8 (UDS 12–4 + full refinement)

Sequence	Video 0 (%)	Video 1 (%)	Video 2 (%)	Video only (%)	Video total (%)	Synthesized (%)
Balloons	0.0000	0.5460	0.7834	0.2955	0.2875	0.2089
Kendo	0.0000	0.7863	0.2575	0.2417	0.2332	0.1295
Newspaper_CC	0.0000	1.3536	1.6544	0.6440	0.6014	0.4292
GT_Fly	0.0000	1.7452	1.9356	0.7626	0.7982	0.4630
Poznan_Hall2	0.0000	2.8343	2.1075	1.1305	1.0715	0.8913
Poznan_Street	0.0000	2.0740	2.8851	0.9051	0.9075	0.6714
Undo_Dancer	0.0000	1.6175	1.4540	0.6705	0.7609	0.5576
Shark	0.0000	0.7848	0.4468	0.3090	0.3368	0.2005
1024×768	0.0000	0.8953	0.8984	0.3937	0.3740	0.2559
1920×1088	0.0000	1.8112	1.7658	0.7555	0.7750	0.5568
Average	0.0000	1.4677	1.4406	0.6199	0.6246	0.4439

References

3D-HEVC Reference Software. Available in: <https://hevc.hhi.fraunhofer.de/svn/svn_3DVCSoftw are/tags/>. Retrieved in: Oct. 2024.

Afonso, V.; Conceição, R.; Saldanha, M.; Braatz, L.; Perleberg, M.; Corrêa, G.; Porto, M.; Agostini, L.; Zatt, B.; Susin, A. Energy-Aware Motion and Disparity Estimation System for 3D-HEVC with Run-Time Adaptive Memory Hierarchy. **Transactions on circuits and systems for video technology (TCSVT)**, v. 29, n. 6, pp. 1878–1892, 2019a.

Afonso, V.; Maich H.; Agostini, L.; Franco, D. Low Cost and High Throughput FME Interpolation for the HEVC Emerging Video Coding Standard. In: IEEE 4th Latin American Symposium on Circuits and Systems, LASCAS, 2013. **Proceedings...** Cusco: Feb. 27 - Mar. 1, 2013.

Afonso, V.; Maich H.; Audibert, L.; Zatt, B.; Porto, M.; Agostini, L.; Susin, A. Hardware Implementation for the HEVC Fractional Motion Estimation Targeting Real-Time and Low-Energy. **Journal of Integrated Circuits and Systems (JICS)**, v.11, n. 2, pp. 106−120, Apr. 2016.

Afonso, V.; Saldanha, M.; Conceição, R.; Perleberg, M.; Porto, M., Zatt, B.; Susin, A.; Agostini, L. Real-time architectures for 3D video coding. **In VLSI Architectures for Future Video Coding**, ch. 6, pp. 191–226. The Institution of Engineering and Technology (IET), England, 2019b.

Afonso, V.; Susin, A.; Audibert, L.; Saldanha, M.; Conceição, R.; Porto, M.; Zatt, B.; Agostini, L. Low-power and high-throughput hardware design for the 3D-HEVC depth intra skip. In: IEEE International Symposium on Circuits and Systems, ISCAS, 2017. **Proceedings...** Baltimore: IEEE, 2017.

Afonso, V.; Susin, A.; Perleberg, M.; Conceição, R.; Correa, G.; Agostini, L.; Zatt, B.; Porto, M. Hardware-Friendly Unidirectional Disparity-Search Algorithm for 3D-HEVC. In: 2018 IEEE International Symposium on Circuits and Systems, ISCAS, 2018. **Proceedings...** Florence: IEEE, 2018.

Agostini, L. **Desenvolvimento de Arquiteturas de Alto Desempenho Dedicadas a Compressão de Vídeo Segundo o Padrão H.264/AVC.** 2007. 172f. Ph.D. Thesis (Doctorate in Computing Science) – Instituto de Informática, UFRGS, Porto Alegre.

Ahmad, W.; Martina, M.; Masera, G. Complexity and Implementation Analysis of synthesized view distortion estimation architecture in 3D High Efficiency Video Coding. In: International Conference on 3D Imaging, IC3D, 2015. **Proceedings...** Liege: 2015.

Alcocer, E.; Gutierrez, R.; Lopez-Granado, O.; Malumbres, M. Design and implementation of an efficient hardware integer motion estimation for an HEVC video encoder. **Journal of Real-Time Image Processing**, v. 16, pp. 547−557, 2019.

AMD. Comparing ASIC gate-equivalent with XU LUTs. Available in: <https://adaptivesupport. amd.com/s/question/0D52E00006hpjEmSAI/comparing-asic-gateequivalent-with-xu-luts?lan guage=en_US> Retrieved in: Oct., 2024.

V. Afonso et al., *Hardware Design for 3D Video Coding*, Synthesis Lectures on Engineering, Science, and Technology, https://doi.org/10.1007/978-3-031-80232-4

211

Amish, F.; Bourennane E. An efficient hardware solution for 3D-HEVC intra-prediction. **Journal of Real-Time Image Processing**, v. 16, pp. 1559−1571, 2019.

Bjontegaard, G. **Calculation of average PSNR differences between RD curves (VCEG-M33)**, Austin, 2001.

Bjontegaard, G. **Improvements of the BD-PSNR model (VCEG-AI11)**, Jul. 2008.

BOSSEN, F. **Common test conditions and software reference configurations (JCTVC-L1100)**, Jan. 2013.

Boyce, J. M.; Doré, R.; Dziembowski, A.; Fleureau, J.; Jung, J.; Kroon, B.; Salahieh, B.; Vadakital, V.K.M.; Yu, L. MPEG Immersive Video Coding Standard. **In Proceedings of the IEEE**, v. 109, n. 9, pp. 1521−1536, Sept. 2021.

Braatz, L.; Zatt, B.; Palomino, D.; Agostini, L.; Porto, M. High-Throughput and Low-Power Integrated Direct/Inverse HEVC Quantization Hardware Design. In: International Symposium on Circuits and Systems, ISCAS, 2018. **Proceedings...** Florence: IEEE, 2018.

Bresenham, J. Algorithm for computer control of a digital plotter. **IBM Systems Journal**, [S.l.], v.4, n.1, p.25–30, 1965.

Bross, B.; Han, W.; Ohm, J.; Sullivan, G.; Wiegand, T. **High Efficiency Video Coding text specification draft 9 (JCTVC-K1003)**, Shangai, 2012.

CADENCE. **Encounter RTL Compiler.** Available in: < https://www.cadence.com/en_US/home/ multimedia.html/content/dam/cadence-www/global/en_US/videos/training/why-you-should- take-encounter-rtl-compiler-cadence-training-course->. Retrieved in: Oct., 2024.

Carroll A.; Heiser, G. The Systems Hacker's Guide to the Galaxy Energy Usage in a Modern Smartphone. In: 4th Asia-Pacific Workshop on Systems, APSys, 2013. **Proceedings...** Singapore: ACM, 2013.

Chen, C.; Huang, C.; Chen, Y.; Chen, L. Level C+ Data Reuse Scheme for Motion Estimation With Corresponding Coding Orders. **Transactions on Circuits and Systems for Video Technology (TCSVT)**, v. 16, n. 4, pp. 553−558, Apr. 2006.

Chen, Y.; Tech, G.; Wegner, K.; Yea, S. **Test Model 11 of 3D-HEVC and MV-HEVC (JCT3V-K1003)**, Geneva, 2015.

Cho, S.; Kim, H.; Kim, H. Y.; Kim, M. Efficient In-Loop Filtering Across Tile Boundaries for Multi-Core HEVC Hardware Decoders With 4 K/8 K- UHD Video Applications. **Transactions on Multimedia (TMM)**, v. 17, n. 6, pp. 778- 791, 2015.

Choi, M.; Chang, I.; Kim, J. High Performance and Hardware Efficient Multiview Video Coding Frame Scheduling Algorithms and Architectures. **Transactions on Circuits and Systems for Video Technology (TCSVT)**, v. 23, n. 8, pp. 1312−1321, Aug. 2013.

Conceição, R.; Avila, G.; Corrêa, G.; Porto, M.; Zatt B.; Agostini, L. Complexity Reduction for 3D-HEVC Depth Map Coding Based On Early Skip and Early DIS Scheme. In: IEEE International Conference on Image Processing, ICIP, 2016. **Proceedings...** Phoenix: IEEE, 2016.

Correa, M.; Zatt, B.; Porto, M.; Agostini, L. High-throughput HEVC intrapicture prediction hardware design targeting UHD 8K videos. In: International Symposium on Circuits and Systems, ISCAS, 2017. **Proceedings...** Baltimore: IEEE, 2017.

Dimenco. Available in: < https://dimenco.com.br/display/>. Retrieved in: Oct. 2024.

Ding, L.; Chen, W.; Tsung, P.; Chuang, T.; Hsiao, P.; Chen, Y.; Chiu, H.; Chien, S.; Chen, L. A 212 MPixels/s 4096 2160p Multiview Video Encoder Chip for 3D/Quad Full HDTV Applications. **IEEE Journal of Solid-State Circuits**, v. 45, n. 1, pp. 46−58, Jan. 2010.

Doan, N.; Kim, T.; Rhee, C.; Lee, H. A hardware-oriented concurrent TZ search algorithm for High-Efficiency Video Coding. **EURASIP Journal on Advances in Signal Processing**, v. 2017, n. 1, Nov. 2017.

EPSON. Moverio BT-200 Smart Glasses. Available in: <https://epson.com/Certified-ReNew/Wea rables/Moverio-BT-200-Smart-Glasses-%28Developer-Version-Only%29/p/V11H560020>. Retrieved in: Oct., 2024.

FANG, C.; CHEN, I.; CHANG, T. A hardware-efficient deblocking filter design for HEVC. In: IEEE International Symposium on Circuits and Systems, ISCAS, 2015. **Proceedings...** Lisbon: IEEE, 2015, pp. 1786–1789.

Fan, Y.; Huang, L.; Hao, B.; Zeng, X. A Hardware-Oriented IME Algorithm for HEVC and Its Hardware Implementation. **Transactions on Circuits and Systems for Video Technology (TCSVT)**, v. 28, n. 8, pp. 2048–2057, 2018.

Geng, J. Three-dimensional display technologies. **Advances in Optics and Photonics**, [S.l.], v. 5, n. 4, p. 456–535, Nov. 2013.

Ghanbari, M. **Standard Codecs:** Image Compression to Advanced Video Coding. United Kingdom: The Institution of Electrical Engineers, 2003.

Goebel, J.; Paim, G.; Agostini, L.; Zatt, B.; Porto, M. An HEVC multi-size DCT hardware with constant throughput and supporting heterogenous CUs. In: International Symposium on Circuits and Systems, ISCAS, 2016. **Proceedings...** Montreal: IEEE, 2016.

Gogoi, S.; Peesapati, R. A hybrid hardware oriented motion estimation algorithm for HEVC/H.265. **Journal of Real-Time Image Processing**, v. 18, pp. 953–966, 2021.

Goldman, M. S.; Litwic, L.; Baumann, O. ULTRA-HD Content Acquisition and Exchange Using HEVC Range Extensions. **SMPTE Motion Imaging Journal**, v. 124, n. 3, p. 28–36, Apr. 2015.

Gonzalez, R.; Woods, R. **Processamento de Imagens Digitais.** São Paulo: Edgard Blücher, 2003.

Gopalakrishna, S.; Hannuksela, M.; Gabbouj, M. Flexible Coding Order for 3D Video Extension of H.265/HEVC. In: IEEE Picture Coding Symposium, PCS, 2013. **Proceedings...** San Jose: IEEE, Dec. 8–11, 2013, pp. 253–256.

Grellert, M. **Computational Effort Analysis and Control in High Efficiency Video Coding.** 2014. 89f. Dissertação de Mestrado. Mestrado em Ciência da Computação – Instituto de Informática, UFRGS, Porto Alegre.

He, G.; Zhou, D.; Li, Y.; Chen, Z.; Zhang T.; Goto, S. High-Throughput Power-Efficient VLSI Architecture of Fractional Motion Estimation for Ultra-HD HEVC Video Encoding. **IEEE Transactions on Very Large Scale Integration (VLSI) Systems**, v. 23, n. 12, pp.3138–3142, Dec. 2015.

Huang, X.; Jia, H.; Cai, B.; Zhu, C.; Liu, J.; Yang, M.; Xie, D.; Gao, W. Fast algorithms and VLSI architecture design for HEVC intra-mode decision. **Journal of Real-Time Image Processing**, v. 12, pp. 285–302, 2016.

Ikai, T.; Tsukuba, T. **Simplification of DMM table derivation (JCT3V-K0042)**, Geneva, 2015.

INTEL. High-Performance FPGA Architecture. Intel® FPGAs and Programmable Devices. Available in: <https://www.intel.com/content/www/us/en/products/details/fpga/stratix.html>. Retrieved in: Oct., 2024:

Intel Realsense Technology. Available in: <https://www.intel.com/content/www/us/en/architecture-and-technology/realsense-overview.html>. Retrieved in: Oct., 2024

ISO/IEC. **Call for Proposals on 3D Video Coding Technology (JTC1/SC29/WG11)**, Geneva, 2011.

ITU-T. **Joint Draft 8.0 on Multiview video coding (JVT-AB204)**, 2008.

ITU-T. **Recommendation H.264: Advanced video coding for generic audiovisual services,** Aug., 2017.

ITU-T. **Recommendation ITU-T H.265: High efficiency video coding,** Apr. 2015.

Jia, L.; Au, O. C.; Tsu, C.; Shi, Y.; Ma, R.; Zhang, H. A diamond search windowbased adaptive search range algorithm. In: IEEE International Conference on Multimedia & Expo Workshops, ICMEW, 2013. **Proceedings...** San Jose: IEEE, 2013.

Jia, L.; Tsui, C.; Au, O.; Zheng, A. A Fast Variable Block Size Motion Estimation Algorithm with Refined Search Range for A Two-layer Data Reuse Scheme. In: IEEE International Symposium on Circuits and Systems, ISCAS, 2015. **Proceedings...** Lisbon: IEEE, May. 24–27, 2015, pp. 1206–1209.

Jiang, C.; Nooshabadi, S. A Scalable Massively Parallel Motion and Disparity Estimation Scheme for Multiview Video Coding. **IEEE Transactions on Circuits and Systems for Video Technology**, v. 26, n. 2, pp. 346–359, Feb. 2016.

Jou, S.; Chang, S.; Chang, T. Fast Motion Estimation Algorithm and Design for Real Time QFHD High Efficiency Video Coding. **IEEE Transactions on Circuits and Systems for Video Technology**, v. 25, n. 9, pp. 1533–1544, 2015.

Kalali, E.; Hamzaoglu, I. Approximate HEVC fractional Interpolation Filters and Their Hardware Implementations. **Transactions on Consumers Electronics (TCE)**, v. 64, n. 3, pp. 285–291, 2018.

Kalali, E.; Ozcan, E.; Yalcinkaya, O.; Hamzaoglu, I. A low energy HEVC inverse transform hardware. **Transactions on Consumer Electronics (TCE)**, v. 60, n. 4, pp. 754–761, 2014.

Kauff, P.; Atzpadin, N.; Fehn, C.; Müller, M.; Schreer, O.; Smolic, A.; Tanger, R. Depth Map Creation and Image-based Rendering for Advanced 3DTV Services Providing Interoperability and Scalability. **Signal Processing: Image Communication (SPIC)**, v. 22, n. 2, Feb. 2007.

Kim, S.; Lee, D.; Sohn, C.; Oh, S. Fast motion estimation for HEVC with adaptive search range decision on CPU and GPU. In: IEEE China Summit & International Conference on Signal and Information Processing, ChinaSIP, 2014. **Proceedings...** Xi'an: IEEE, July 9–13, 2014, pp. 349–353.

Kim, M.; Ling, N.; Song, L. Fast single depth intra mode decision for depth map coding in 3D-HEVC. In: IEEE International Conference on Multimedia & Expo Workshops, ICMEW, 2015. **Proceedings...** Turin: IEEE, 2015a.

Kim, D.; Moon, J.; Lee, S. Hardware implementation of HEVC CABAC encoder. In: International SoC Design Conference, ISOCC, 2015. **Proceedings...** Gyungju: IEEE, 2015b.

Lainema, J. Intra-Picture Prediction in HEVC. In: SZE, V.; BUDAGAVI, M.; SULLIVAN, G. (Ed.). High efficiency video coding (HEVC). [S.l.]: Springer, 2014. p.91–112.

Lee, B.; Kim, M. A CU-Level Rate and Distortion Estimation Scheme for RDO of Hardware-Friendly HEVC Encoders Using Low-Complexity Integer DCTs. **Transactions on Image Processing (TIP)**, v. 25, n. 8, pp. 3787–3800, 2016.

Lee, J.; Park, M.; Choi, B.; Cho, Y.; Kim, C. **3D-CE5 related: Segment-wise depth intra mode coding (JCT3V-G0102)**, San José, 2014.

Lee, J.; Park, M.; Kim, C. **3D-CE1: Depth intra skip (DIS) mode (JCT3V-K0033)**, Geneva, 2015.

Lee, T.; Chan, Y.; Siu, W. Adaptive Search Range for HEVC Motion Estimation Based on Depth Information. **IEEE Transactions on Circuits and Systems for Video Technology**, [S.l.], v. 27, n. 10, pp. 2216–2230, Oct. 2017.

LG. Available in: <https://www.lg.com/br/telemoveis-lg/smartphones/p920/?srsltid=AfmBOopS6 uFqC-p8cMvuUB4LqYFN2kLy_rATtgAFYW8ItF6ty8UzqKFW>. Retrieved in: Oct. 2024.

LI, L. **Time-of-Flight Camera – An Introduction (Texas Instruments – Technical White Paper)**, Dallas, 2014a.

Li, X.; Wang, R.; Wang, W.; Wang Z.; Dong, S. Fast motion estimation methods for HEVC. In: IEEE International Symposium on Broadband Multimedia Systems and Broadcasting, 2014. **Proceedings...** Beijing: IEEE, 2014b.

Lifewire. Available in: < https://www.lifewire.com/why-3d-tv-died-4126776>. Retrieved in: Sep. 2024.

Liu, W.; Li, J.; Cho, Y. B. A novel architecture for parallel multi-view HEVC decoder on mobile device. **EURASIP Journal on Image and Video Processing**, Mar. 2017.

Lung, Cy.; Shen, Ca. Design and implementation of a highly efficient fractional motion estimation for the HEVC encoder. **Journal of Real-Time Image Processing**, v. 16, pp. 1541–1557, 2019.

Mccann, K.; ROSEWARNE, C.; BROSS, B.; NACCARI, M.; SHARMAN, K.; SULLIVAN, G. **High Efficiency Video Coding (HEVC) Test Model 16 (HM 16) Improved Encoder Description (JCTVC-S1002)**, Strasbourg, 2014.

Merkle, P.; Smolic, A.; Müller, K.; Wiegand, T. Efficient Prediction Structures for Multiview Video Coding. **IEEE Transactions on Circuits and Systems for Video Technology**, [S.l.], v. 17, n. 11, p. 1461–1473, nov. 2007.

Micron Technology Inc., **272b: x64 Mobile LPDDR4 SDRAM Features**. MT53B384M64D4, 2014.

Microsoft Hololens Technology. Available in: <https://www.microsoft.com/en-us/hololens>. Retrieved in: Oct., 2024.

Min, B.; Xu, Z.; Cheung, R. A Fully Pipelined Hardware Architecture for Intra Prediction of HEVC. **Transactions on Circuits and Systems for Video Technology (TCSVT)**, v.27, n. 12, Dec. 2017.

Müller, K.; Schwarz, H.; Marpe, D.; Bartnik, C.; Bosse, S.; Brust, H.; Hinz, T.; Lakshman, H.; Merkle, P.; Rhee, F. H.; Tech, G.; Winken, M.; Wiegand, T. 3D High-Efficiency Video Coding for Multi-View Video and Depth Data. **IEEE Transactions on Image Processing**, [S.l.], v. 22, n. 9, p. 3366–3378, sep. 2013.

Müller, K.; Vetro, A. **Common Test Conditions of 3DV Core Experiments (JCT3V-G1100)**, San José, 2014.

Nangate. Open Cell and Free PDK Libraries. Available in: < https://si2.org/open-cell-and-free-pdk-libraries/>. Retrieved in: Oct., 2024.

Nintendo. Available in: < https://www.nintendo.com/pt-pt/Consolas-e-acessorios/Familia-Nintendo-3DS/New-Nintendo-3DS-XL/New-Nintendo-3DS-XL-955921.html>. Retrieved in: Oct. 2024.

NVSIM. Available in: < https://github.com/SEAL-UCSB/NVSim>. Retrieved in: Oct., 2024.

Ohm, J.-R. **Multimedia Signal Coding and Transmission.** Aachen: Springer , 2015.

PALOMINO, D.; SAMPAIO, F.; AGOSTINI, L.; BAMPI, S.; SUSIN, S. A memory aware multiplierless VLSI architecture for the complete Intra Prediction of the HEVC emerging standard. In: International Conference on Image Processing, ICIP, 2012. **Proceedings...** Orlando: IEEE, Oct. 2012.

Panasonic Hdc-Sdt750k. Available in: < https://www.panasonic.com/content/dam/Panasonic/support_manual/Camcorder_Digital/English/HDC-SDT750_vqt3b52_MAIN_OI.pdf>. Retrieved Oct., 2024.

Pastuszak, G.; Abramowski, A. Algorithm and Architecture Design of the H.265/HEVC Intra Encoder. **Transactions on Circuits and Systems for Video Technology (TCSVT)**, v. 26, n. 1, pp. 210–222, 2016a.

Pastuszak G.; Trochimiuk, M. Algorithm and architecture design of the motion estimation for the H.265/HEVC 4K-UHD encoder. **Journal of Real-Time Image Processing**, v. 12, n. 2, pp. 517–529, Aug. 2016b.

Pereira, F.; Da Silva, E. A. B. Efficient Plenoptic Imaging Representation: Why Do We Need It?. In: IEEE International Conference on Multimedia and Expo, ICME, 2016. **Proceedings...** Seattle: IEEE, 2016.

Perleberg, M.; Afonso, V.; Conceição, R.; Susin, A.; Agostini, L.; Zatt, B.; Porto, M. Energy and Rate-Aware Design for HEVC Motion Estimation Based on Pareto Efficiency. **Journal of Integrated Circuits and Systems (JICS)**, v. 13, pp. 1−12, 2018.

Perleberg, M.; Afonso, V.; Conceição, R.; Susin, A.; Agostini, L.; Zatt, B.; Porto, M. High-Throughput Hardware Design for 3D-HEVC Disparity Estimation. **IEEE Design & Test**, v. 37, n. 3, pp. 22−29, June 2020a.

Perleberg, M.R.; Afonso, V.; Borges, V.A.; Zatt, B.; Agostini, L.V., Porto, M. Quality-power configurable flexible coding order hardware design for real-time 3D-HEVC intra-frame prediction.
Journal of Real-Time Image Processing, v. 19, pp. 969–984, 2022.

Perleberg, M.; Borges, V.; Afonso, V.; Palomino, D.; Agostini, L.; Porto, M. 6WR: A Hardware
Friendly 3D-HEVC DMM-1 Algorithm and its Energy-Aware and High-Throughput Design.
IEEE Transactions on Circuits and Systems II: Express Briefs. v. 67, n. 5, pp. 836–840,
May 2020b.

Perleberg, M.; Goebel, J.; Melo, M.; Agostini, L.; Zatt, B.; Porto, M. ASIC power-estimation accuracy evaluation: A case study using video-coding architectures. In: IEEE 9th Latin American
Symposium on Circuits & Systems, LASCAS, 2018. **Proceedings...** Puerto Vallarta: IEEE, 2018.

Ramos, F.; Goebel, J.; Zatt, B.; Porto, M.; Bampi, S. Low-power hardware design for the HEVC
binary arithmetic encoder targeting 8K videos. In: Symposium on Integrated Circuits and Systems Design, SBCCI, 2016. **Proceedings...** Belo Horizonte: ACM/IEEE, sep. 2016.

Ramos, F.; Zatt, B.; Porto, M.; Bampi, S. High-Throughput Binary Arithmetic Encoder using
Multiple-Bypass Bins Processing for HEVC CABAC. In: IEEE International Symposium on
Circuits and Systems, ISCAS, 2018. **Proceedings...** Florence: IEEE, 2018.

Richardson, I. **H.264 and MPEG-4 Video Compression**: Video Coding for Next-Generation
Multimedia. Chichester: John Wiley and Sons, 2003.

Richardson, I. **Video Codec Design**: Developing Image and Video Compression Systems. Chichester: John Wiley and Sons, 2002.

Sampaio, F.; Zatt, B.; Shafique, M.; Agostini, L.; Bampi, S.; Henkel, J. Energy-Efficient Memory
Hierarchy for Motion and Disparity Estimation in Multiview Video Coding. In: Design, Automation & Test in Europe Conf. & Exhibition, DATE, 2013. **Proceedings...** Grenoble: Mar. 18–22,
2013, pp. 665–670.

Samsung. **Galaxy SIII**. Available in: < https://www.samsung.com/ae/support/model/GT-I9300M
BDXSG/>. Retrieved in: Oct. 2024.

Sanchez, G.; Saldanha, M.; Fernandes, R.; Cataldo, R.; Agostini, L.; Marcon, C. 3D-HEVC Bipar-
tition Modes Encoder and Decoder Design Targeting High-Resolution Videos. **IEEE Transactions on Circuits and Systems I: Regular Papers**, Ago. 2019.

Sanchez, G.; Agostini, L.; Marcon, C. Complexity reduction by modes reduction in RD-list for
intra-frame prediction in 3D-HEVC depth maps. In: International Symposium on Circuits and
Systems, ISCAS, 2017. **Proceedings...** Baltimore: IEEE, 2017a.

Sanchez, G.; Marcon, C.; Agostini, L. High Efficient Architecture for 3D-HEVC DMM-1 Decoder
Targeting 1080p Videos. In: International Symposium on Circuits and Systems, ISCAS, 2018.
Proceedings... Florence: IEEE, 2018a.

Sanchez, G.; Marcon, C.; Agostini, L. Real-time scalable hardware architecture for 3D-HEVC bipar-
tition modes. **Journal of Real-Time Image Processing**, v.13, n.1, pp. 71–83, jun. 2016.

Sanchez, G.; Saldanha, M.; Balota, G.; Zatt, B.; Porto, M.; Agostini, L. Complexity reduction for
3D-HEVC depth maps intra-frame prediction using simplified edge detector algorithm. In: IEEE
International Conference on Image Processing, ICIP, 2014. **Proceedings...** Paris: Oct 27–30.
2014a, pp. 3209–3213.

Sanchez, G.; Saldanha, M.; Balota, G.; Zatt, B.; Porto, M.; Agostini, L. DMMFast: a complexity reduction scheme for three-dimensional high-efficiency video coding intraframe depth map
coding. **Journal of Electronic Imaging**, v. 24, n. 2, p. 1–15, apr. 2015a.

Sanchez, G.; Saldanha, M.; Porto, M.; Zatt, B.; Agostini, L. Real-time simplified edge detector architecture for 3D-HEVC depth maps coding. In: International Conference on Electronics, Circuits
and Systems, ICECS, 2017. **Proceedings...** Monte Carlo: IEEE, 2017b.

Sanchez, G.; Silveira, J.; Agostini, L.; Marcon, C. Performance Analysis of Depth Intra Coding in 3D-HEVC. **Transactions on circuits and systems for video technology (TCSVT)**, Early Access, aug. 2018b.

Sanchez, G.; Zatt, B.; Porto, M.; Agostini, L. A Real-Time 5-Views HD 1080p Architecture for 3D-HEVC Depth Modeling Mode 4. In: ACM/IEEE Symposium on Integrated Circuits and Systems Design, SBCCI, 2014. **Proceedings...** Aracajú: ACM/IEEE, 2014b. p. 1–6.

Sanchez, G.; Zatt, B.; Porto, M.; Agostini, L. Hardware-friendly HEVC motion estimation: new algorithms and efficient VLSI designs targeting high definition videos. **Analog Integrated Circuits and Signal Processing**, v. 82, n. 1, pp. 135−146, 2015b.

Seidel, I.; Bräscher, A. B.; Güntzel, J. L.; Agostini, L. Energy-efficient SATD for beyond HEVC. In: IEEE International Symposium on Circuits and Systems, ISCAS, 2016. **Proceedings...** Montreal: ACM/IEEE, 2016.

Shen, W.; Fan, Y.; Bai, Y.; Huang, L.; Shang, Q.; Liu, C.; Zeng, X. A Combined Deblocking Filter and SAO Hardware Architecture for HEVC. **Transactions on Multimedia (TMM)**, v. 18, n. 6, pp. 1022−1033, 2016

Shi, Lz.; Gao, X.; Yang, X.; Chen, Z.; Zheng, M. Algorithm optimization and hardware implementation for Merge mode in HEVC. **Journal of Real-Time Image Processing**, v. 17, pp. 623–630, 2020.

Shi, Lz.; Yan, D.; Hong, X.; Huang, B.; Yang, X. Algorithm optimisation and hardware implementation of interprediction mode decision. **Journal of Real-Time Image Processing**, v. 18, pp. 593–601, 2021.

Smolic, A.; Kauff, P.; Knorr, S.; Hornung, A.; Kunter, M.; Müller, M.; Lang, M. Three-Dimensional Video Postproduction and Processing. **Proceedings of the IEEE**, [S.l.], v. 99, n. 4, p. 607–625, apr. 2011.

Song, C.; Ju, L.; Jia, Z. Hybrid Scratchpad and Cache Memory Management for Energy-Efficient Parallel HEVC Encoding. In: 33rd IEEE In: International Conference on Computer Design, ICCD, 2015. **Proceedings...** New York: IEEE, Oct. 18–21, 2015, pp. 712–719.

SONY. Available in: <http://www.sonyrumors.net/wp-content/uploads/2011/06/3D-Movie-Tim eline-Medium.jpg?bdc16b>. Retrieved in: Oct. 2024.

STEREOLABS ZED. Available in: < https://www.stereolabs.com/en-br/products/zed-2>. Retrieved in: Oct. 2024.

STRUCTURE INC. SENSOR. Available in: <https://structure.io/>. Retrieved in: Oct. 2024.

Sullivan, G.; Ohm, J.; Han, W.; Wiegand, T. Overview of the High Efficiency Video Coding (HEVC) Standard. **IEEE Transactions on Circuits and Systems for Video Technology**, [S.l.], v. 22, n. 12, p. 1649–1668. sep. 2012.

Sullivan, G.; Boyce, J.; Chen, Y.; Ohm, J.; Segall, C.; Vetro, A. Standardized extensions of High Efficiency Video Coding (HEVC). **Journal of Selected Topics in Signal Processing**, v. 7, n. 6, pp. 1001−1016, 2013.

Tech, G.; Chen, Y.; Müller, K.; Ohm, J.; Vetro, A.; Wang, Y. Overview of the Multiview and 3D extensions of High Efficiency Video Coding. **IEEE Transactions on Circuits and Systems for Video Technology (TCSVT)**, [S.l.], v. 26, n. 1, p. 35–49, Jan. 2016.

Tech, G.; Wegner, K.; Chen, Y.; Yea, S. **3D-HEVC Draft Text 7 (JCT3V-K1001)**, Geneva, 2015.

Toshiba. Available in: <https://www.manuals.ca/toshiba/55zl2g/manual?p=1>. Retrieved in: Oct. 2024.

Tseng, Y.; Shen, C. The Design and Implementation of a Highly Efficient Motion Estimation Engine for HEVC Systems. In: IEEE International Symposium on Circuits and Systems, ISCAS, 2019. **Proceedings...** Sapporo: IEEE, 2019.

TSMC. 40nm Technology. Available in: https://www.tsmc.com/english/dedicatedFoundry/techno logy/logic/l_40nm. Retrieved in: Oct. 2024.

Tsung, P.; Chen, W.; Ding, L.; Chien, S.; Chen, L. Cache-based integer motion/disparity estimation for quad-HD H.264/AVC and HD multiview video coding. In: IEEE International Conference on Acoustics, Speech and Signal Processing, ICASSP, 2009. **Proceedings...** Taipei: IEEE, Apr. 19–24, 2009, pp. 2013–2016.

Ücker, M.; Afonso, V.; Audibert, L.; Susin, A.; Zatt, B.; Porto, M.; Agostini, L. Low-Power and High-Throughput Architecture for 3D-HEVC Depth Modeling Mode 4. In: 31st Symposium on Integrated Circuits and Systems Design, SBCCI, 2018. **Proceedings...** Bento Goncalves: IEEE, 2018.

Ücker, M.; Afonso, V.; Saldanha, M.; Audibert, L.; Conceição, R.; Susin, A.; Zatt, B.; Porto, M.; Agostini, L. High-Throughput Hardware for 3D-HEVC Depth-Map Intra Prediction. IEEE Design & Test, v. 37, n. 3, pp. 7–14, June 2020.

Vetro, A.; Pandit, P.; Kimata, H.; Smolic, A.; Wang, Y-K. **Joint Draft 8.0 on Multiview video coding. (JVT-AB204)**, Hannover, 2008.

VITECH. 3D-HEVC Memory Simulator. Available in: <https://wp.ufpel.edu.br/vitech/en/downloads/>. Retrieved in: Oct. 2024.

Wang, B.; Zhou, J.; Kim, T. SRAM devices and circuits optimization toward energy efficiency in multi-Vth CMOS. **Microelectronics Journal**, v. 46, n. 3, pp. 265–272, mar. 2015.

Xu, K.; Huang, B.; Liu, X.; Tu, X.; Wu, Z.; Yan, Z.; Liu, P.; Han, B; Li, Y. A Low-power Pyramid Motion Estimation Engine for 4K@30fps Realtime HEVC Video Encoding. In: IEEE International Symposium on Circuits and Systems, ISCAS, 2018. **Proceedings...** Florence: IEEE, 2018.

Zatt, B. **Energy-Efficient Algorithms and Architectures for Multiview Video Coding**. 2012. 236 f. Ph.D. Thesis (Doctorate on Microelectronics) – Instituto de Informática, Universidade Federal do Rio Grande do Sul, Porto Alegre, 2012.

Zatt, B.; Shafique, M.; Bampi, S.; Henkel, J. **3D Video Coding for Embedded Devices**: Energy Efficient Algorithms and Architectures. Springer Science, New York, 2013.

Zatt, B.; Shafique, M.; Bampi, S.; Henkel, J. A low-power memory architecture with application-aware power management for motion & disparity estimation in Multiview Video Coding. In: IEEE/ACM Int'l Conf. Computer-Aided Design, ICCAD, 2011. **Proceedings...** San Jose: IEEE/ACM, Nov. 7–10, 2011a, pp. 40–47.

Zatt, B.; Shafique, M.; Sampaio, F.; Agostini, L.; Bampi, S.; Henkel, J. Run-time Adaptive Energy-Aware Motion and Disparity Estimation in Multiview Video Coding. In: 48th ACM/EDAC/IEEE Design Automation Conference, DAC, 2011. **Proceedings...** New York: ACM/EDAC/IEEE, Jun. 5–9, 2011b, pp. 1026–1031.

Zhang, Y.; Lu, C. A Highly-Parallel Hardware Architecture of Table-Based CABAC Bit Rate Estimator in HEVC intra encoder. **Transactions on Circuits and Systems for Video Technology (TCSVT)**, Early Access, apr. 2018.

Zhang, X.; Zhang, K.; An, J.; Huang, H.; Lin, J.-L.; Lei, S. **On Lookup Table Size Reduction for DMM1 (JCT3V-J0035)**, Strasbourg, 2014.

Zhang, Y.; Lu, C. Efficient Algorithm Adaptations and Fully Parallel Hardware Architecture of H.265/HEVC Intra Encoder. **IEEE Transactions on Circuits and Systems for Video Technology (TCSVT)**, v. 29, n. 11, pp. 3415–3429, 2019.

Zhao, L.; Zhang, L.; Ma, S.; Zhao, D. Fast mode decision algorithm for intra prediction in HEVC. In: Visual Communications and Image Processing, VCIP, 2011. **Proceedings...** Tainan: IEEE, 2011, pp. 1–4.

Zhou, W.; Zhang, J.; Zhou, X.; Liu, Z.; Liu, X.; A High-Throughput and Multi-Parallel VLSI Architecture for HEVC Deblocking Filter. **Transactions on Multimedia (TMM)**, v. 18, n. 6, pp. 1034–1047, 2016.